SAS/GIS® 9.2
Spatial Data and Procedure Guide

SAS® Documentation

The correct bibliographic citation for this manual is as follows: SAS Institute Inc. 2009. *SAS/GIS® 9.2: Spatial Data and Procedure Guide*. Cary, NC: SAS Institute Inc.

SAS/GIS® 9.2: Spatial Data and Procedure Guide

ISBN 978-1-59994-889-8

SAS Institute Inc., SAS Campus Drive, Cary, North Carolina 27513.

1st electronic book, February 2009

1st printing, March 2009

Contents

PART 1 Usage and Reference 1

PART 1

Usage and Reference

CHAPTER 1

Overview of SAS/GIS Software

Introduction to Geographic Information Systems

SAS/GIS software provides an interactive geographic information system within SAS. A *geographic information system* (GIS) is a tool for organizing and analyzing data that can be referenced spatially, that is, data that can be tied to physical locations. Many types of data have a spatial aspect, including demographics, marketing surveys, customer addresses, and epidemiological studies. A GIS helps you analyze your data in the context of location.

For example, if you need to evaluate population data for census tracts, you could view the information in tabular format. However, consider how much easier and more effective it would be to view the demographic information in the context of the geography of the tracts as in Figure 1.1 on page 4. When viewing information that has a spatial component, you might find it easier to recognize relationships and trends in your data if you view the information in a spatial context.

Figure 1.1 Evaluating Spatially Referenced Data

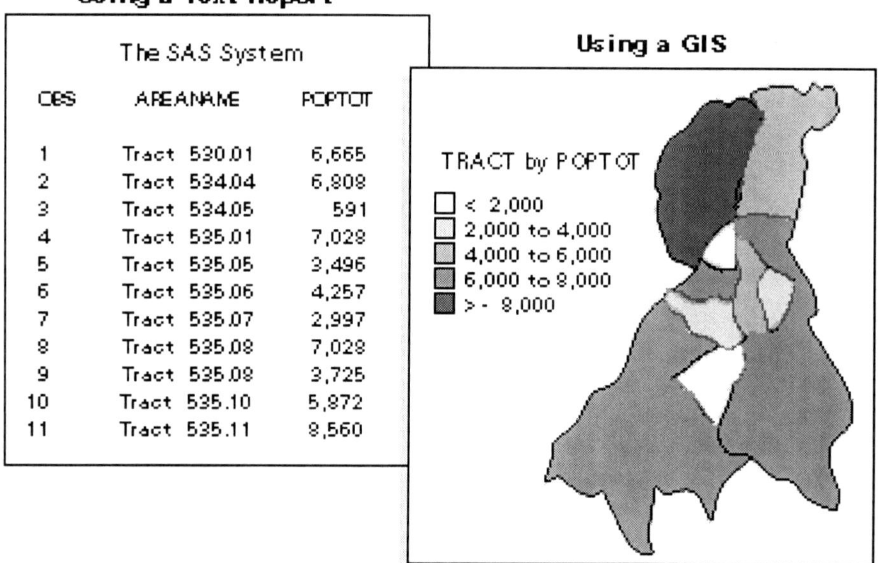

SAS/GIS software enables you to do more than simply view your data in its spatial context. It also enables you to interact with the data by selecting features and performing actions that are based on your selections. SAS/GIS software draws on the capabilities of SAS and enables you to access, manage, analyze, and present your data easily.

Features of SAS Software

SAS provides a powerful programming language with components called procedures that allow you to perform many different kinds of analysis and data management functions, as well as produce many different types of text-based and graphical presentation output. Combined with other features, the SAS language and its procedures make an immense variety of applications possible, including the following examples:

- Access raw data files and data in external databases and database management systems.
- Manage data using tools for data entry, editing, retrieval, formatting, and conversion.
- Analyze data using descriptive statistics, multivariate techniques, forecasting and modeling, and linear programming.
- Present data using reports and business and statistical graphics.

SAS is also portable across computing environments. SAS applications function the same and produce the same results regardless of the operating environment on which you are running SAS to process your data. However, some features, such as interactive windows, are not supported on all platforms.

For more information about SAS, please refer to *SAS Language Reference: Concepts*.

Data in SAS/GIS Applications

SAS/GIS software uses two basic types of data:

spatial data
: contains the coordinates and identifying information that describes the map features such as streets, rivers, and railroads.

attribute data
: is the information that you want to use for analysis or presentation. This information must be spatial in nature. Sales figures for each of your store locations, population data for each county, and total income for each household in a region are examples of information that is spatial in nature because the information applies to a specific geographic feature.

For example, the U.S. Census Bureau distributes both types of data:

TIGER Line files
: contain spatial information that you can use to build maps.

Summary Tape files
: contain population and other demographic information that you can link to the map features.

Attribute data provides the information that you want to analyze, and spatial data provides the context in which you want to analyze it. For example, consider the SAS/GIS map shown in the following display. Spatial data provides the boundaries for the map areas, and attribute data provides the population information that is used to color the map areas.

Display 1.1 Spatial and Attribute Data in SAS/GIS Maps

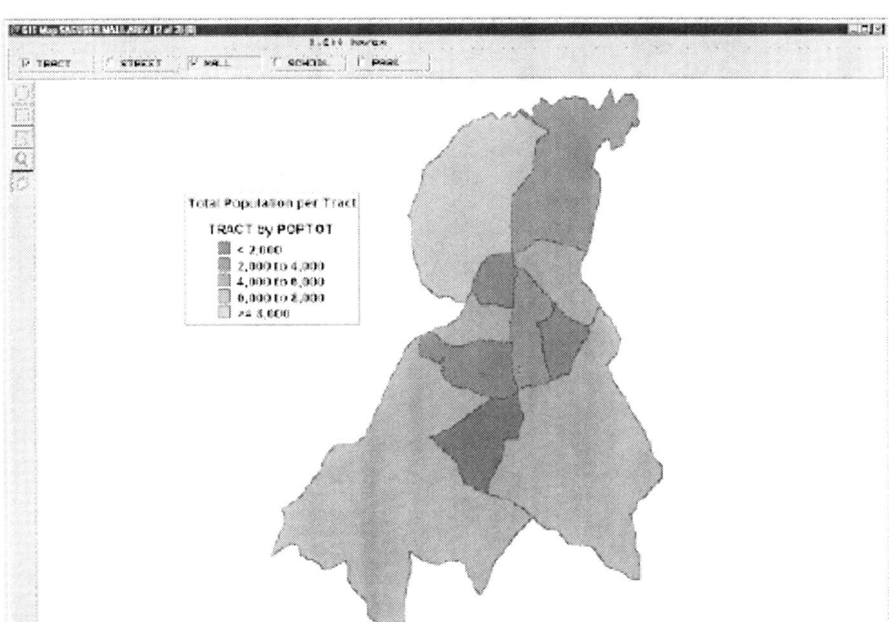

Spatial Data

Spatial data contains the coordinates and identifying information that are necessary to draw maps. For SAS/GIS software, spatial data is stored in SAS/GIS spatial databases, which consist of collections of SAS data sets and SAS catalog entries. The primary method for creating a SAS/GIS spatial database is through the SAS/GIS Import facility, either in batch or in interactive mode. You can also use the GIS procedure to create, modify, and manage the catalog entries in a spatial database.

Spatial Data Layers

Features in the spatial data are organized into layers. A *layer* is a collection of all the features in the map that share some common characteristic. The various physical aspects of the map—political boundaries, roads, railroads, waterways, and so on—are assigned to layers according to their common spatial data values. Some features can appear in multiple layers. For example, a street can also be a ZIP code boundary and a city boundary line. The street could appear in three layers: one containing the streets, one containing the ZIP code boundaries, and one containing the city boundaries.

Three types of layers can be represented in SAS/GIS maps: points, lines, and areas. For example, the collection of all the points in a map that represent park locations can be organized into a point layer for parks, the collection of all the lines in a map that represent streets can be organized into a line layer for streets, and the collection of all the areas that represent census tracts can be organized into an area layer for tracts. When the various layers are overlaid, they form a map, as shown in the following figure.

Figure 1.2 Layers Forming a SAS/GIS Map

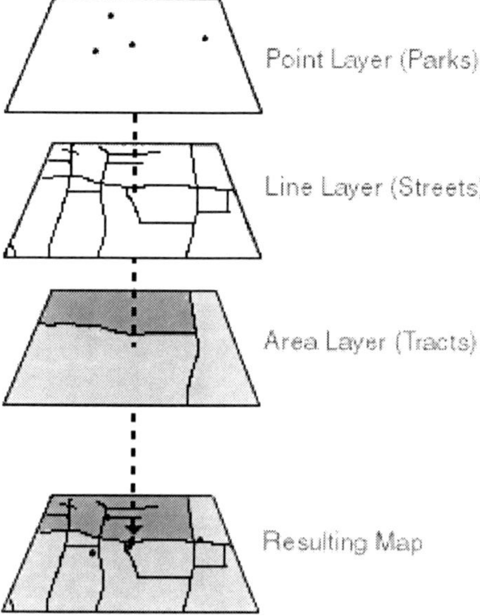

A layer can be displayed as either static or thematic. When a layer is displayed as static, it uses the same graphical characteristics (color, line, width, and so on) for all features in that layer. For example, a street layer could use the same color and line style to display all the streets. When a layer is displayed as thematic, it uses different graphical characteristics to classify the features in that layer. For example, a theme representing sales regions could use different colors to show the quarterly sales performance of each region. A theme in a layer representing highways could use different line widths to show the classes of roads. A layer can have multiple themes stored in it, and you can easily change which theme is currently displayed.

Spatial Data Coverages

In SAS/GIS software, maps display only the portion of the spatial data that falls within a given coverage. A *coverage* defines a subset of the spatial data that is available to a map. The coverage can include all the spatial data in the database, or only selected portions. For example, a spatial database might contain geographic data for an entire country, but a coverage might restrict the portion that is available for a given map to only one region. You can define more than one coverage for each spatial database, although a map uses only one coverage at a time.

Spatial Data Composites

Most operations in SAS/GIS software use composites of spatial data variables rather than the actual spatial data variables themselves. *Composites* identify the relationships and purpose of the variables in the spatial data.

For example, if the spatial data has the variables STATEL and STATER that contain the state ID codes for the left and right sides of each feature, then the spatial database could define a composite named STATE that identifies the relationship between these variables and specifies that they delineate state areas in the map. You would use the STATE composite, rather than the actual STATEL and STATER variables, to link state areas in the map to attribute data for the corresponding state.

See Appendix 2, “Details of SAS/GIS Spatial Databases,” on page 161 for more information about the structure of SAS/GIS spatial databases.

Attribute Data

The second type of data that is used in a GIS is attribute data. In SAS/GIS software, your attribute data must be stored in either a SAS data set or a SAS view. SAS views allow you to transparently access data in other formats. For example, you can create a SAS/ACCESS view to access data in a database such as DB2. A DATA step view or an SQL view also allows you to access an external file, or any other type of data from which you can create a SAS view. Once your attribute data is accessible either as a SAS data set or through a SAS view, it can be linked to your spatial data for use in labeling, analysis, or theming. For example, your spatial data might represent a county and contain information for city boundaries, census tract boundaries, streets, and so on. An attribute data set with population information for each census tract can be linked to a map using the corresponding tract composite in the spatial data.

Some of the ways in which you can use attribute data in SAS/GIS software include the following:

- Use values in your attribute data as labels. For example, you could use attribute data containing population data to provide the text of labels for census tracts.
- Use the values in your attribute data as themes for layers. For example, you could use attribute data containing average household income data as a theme for a census tract layer.

  See Chapter 5, “Customizing Maps,” in *SAS/GIS Software: Usage and Reference, Version 6* for more information about assigning themes to map layers.
- Define actions that display or manipulate the attribute data when features are selected in the map. This way, you can explore your attribute data interactively rather than simply view static results. The actions can range from simple, such as displaying observations from an attribute data set that relate to features in the map, to complex, such as submitting a procedure from SAS/STAT software to perform a statistical analysis.

You can define the following actions for your attribute data:

- Display observations from attribute data sets that relate to selected map features.
- Open additional maps that relate to selected map features.
- Display images that relate to selected map features.
- Interactively subset attribute data sets according to a subset of selected map features.
- Submit SAS programs.
- Issue SAS commands.
- Issue host commands.
- Display and edit information for the selected map features.
- Organize area features into groups that are based on your attribute data.

See Chapter 4, "Performing Actions for Selected Map Features" in *SAS / GIS Software: Usage and Reference, Version 6* for more information on defining and performing actions.

Designing a SAS/GIS Spatial Database

One of the first steps in a SAS/GIS project is determining the design of your SAS/GIS spatial database. The database will contain the following types of information:

Table 1.1 SAS/GIS Data Types

Type of Data	Database Contents
spatial	all of the spatial data that the user wants to see
attribute	all of the associated attribute data that the user needs to use for analysis or presentation purposes

Before you begin creating the spatial database, you should draw up an overview of the system goals and data requirements. The time you spend designing your database initially will save you time and expenses later in the project. A well-designed database is easier to maintain and document, and you can extend it for future GIS projects.

Use the following guidelines when determining the information you want to include in a database:

Table 1.2 SAS/GIS Spatial Database Guidelines

If you want to determine...	Then
project objective	**1** Identify the initial objective of the project and its ultimate goal. **2** Consider any requirements that might have been imposed on it. **3** Determine the feasibility of initial implementation and, as best as possible, the impact of any future demands.
attribute data	**1** Identify the attribute data that is necessary to illustrate the project objectives. **2** Determine whether you have this data or can obtain it.
spatial data	**1** Identify the spatial features that you need to link with your attribute data, for example, states, cities, rivers, roads, railroads, airports, and so on. **2** Determine whether you have this data or can obtain it.

Once you have determined a preliminary list of the data that you will need, use the additional factors in the following sections to help you evaluate and refine your list.

Enable Linking between Spatial Features and Attribute Data

To use attribute data for map actions, themes, or labeling, the attribute data set must contain the same identification information as the spatial feature that it describes so that you can link between them. For example, if your attribute data has Sales Revenue for stores, and Store ID Numbers, you probably want to include the actual location in longitude and latitude for each Store ID Number on your spatial data list. You can then place a marker at the store location and also visualize and analyze the corresponding attribute data for each store.

Use No More Details Than You Need

Use only the data you need for your project. For example, if you have store locations that request the customer ZIP code at the cash register, you should not assume that you need ZIP code boundaries on your map. ZIP code boundaries might be far too small for your purposes if you have stores nationwide. You might decide instead that the three-digit ZIP code boundaries provide fewer, yet more appropriately sized, areas for your analysis. You can summarize your attribute data to the three-digit ZIP code level and use it for your analysis, reducing both the amount of spatial data and attribute data that you need. As long as it is appropriate for your analysis, decreasing the amount of required spatial and attribute data reduces storage space and improves performance. Reducing the level of detail in the spatial data also saves money if you have to purchase the data.

Ensure a Common Level of Spatial and Attribute Data

If you plan to summarize your attribute data to a matching level of your spatial data, make sure that the two types of data have a common level that you can use. For example, ZIP code boundaries can cross not only county boundaries, but also state boundaries, so there is usually not a one-to-one correspondence between ZIP codes and states or counties. If the only information that ties your attribute data to your spatial data is ZIP codes, you will have difficulties using your ZIP code level attribute data if you include only state or county boundaries in your spatial data.

For specific, smaller areas of the country, a one-to-one correspondence might exist that will allow you to summarize your attribute data to a higher level. However, ZIP codes can change frequently, and this correspondence might be lost. Also, because ZIP codes change, you must be able to account for these changes when performing a historical analysis. For example, if you are comparing sales in a specific ZIP code area over a ten-year period, make sure that the area remained constant during that period. The same is true for other spatial data.

Using the SAS/GIS Interface

Starting SAS/GIS Software

Use the following steps to start a SAS/GIS software session:

1 Open a SAS session.
2 From the SAS menu bar, select **Solutions ▸ Analysis ▸ Geographic Information System**

Or type `GIS` in the SAS Command Box or on any SAS command line.

Using Dialog Box Elements

In most places where you must supply a value in a SAS/GIS window, you will see a pull-out arrow, a drop-down arrow, or both, presented in conjunction with text boxes, as shown in the following display.

Display 1.2 Typical Dialog Box Elements

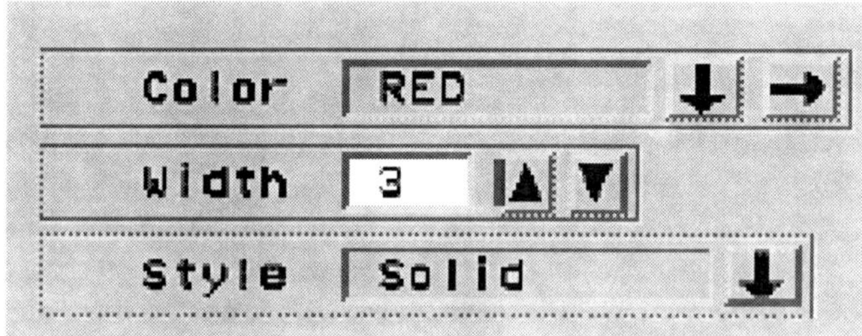

Clicking a drop-down arrow displays a list of valid choices for the option. Display 1.3 on page 11 shows the list that is displayed by clicking the drop-down arrow for the `Style` field in Display 1.2 on page 10.

Display 1.3 List Displayed by the Style Drop-down Arrow

Clicking a pull-out arrow opens a new window in which you can interactively select appropriate values. The following display shows the window that is opened by clicking the pull-out arrow for the `Color` field in Display 1.2 on page 10.

Display 1.4 Window That Is Opened by the Color Pull-out Arrow

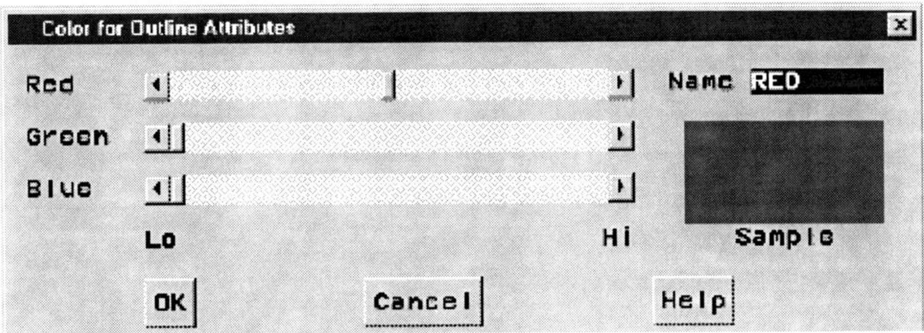

Selecting Maps and SAS Data Sets

Whenever you need to specify the name of a SAS data set or SAS catalog entry, SAS/GIS software opens an Open window like the one shown in the following display.

Display 1.5 Typical Open Window

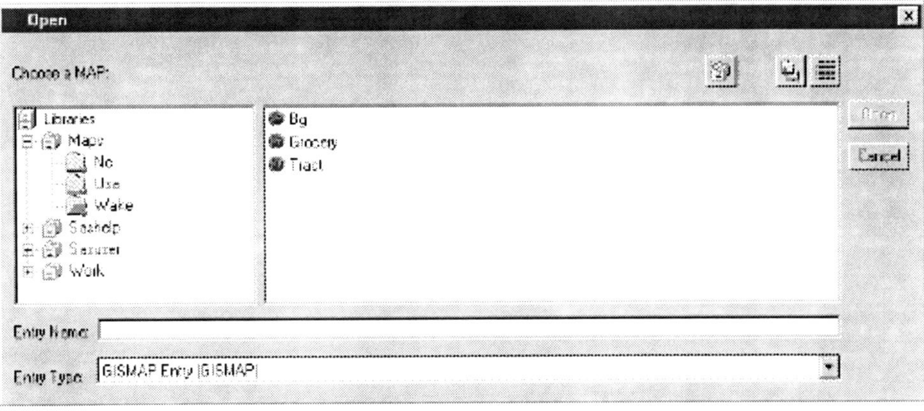

The window provides an intuitive way to find the SAS data set or catalog entry that you need. It also makes it impossible to enter an invalid name, since only those choices that are appropriate for the operation you are performing are presented for selection.

To select each level of the SAS name from the tree view, double-click your choice. Once you make a selection, the list of available choices for the next level of the name is displayed.

Accessing the SAS/GIS Tutorial

This book does not attempt to cover all of the fundamentals of using SAS/GIS software. For an introduction to the basic tasks that you can perform, see the online tutorial that is included in SAS/GIS software. To start the tutorial, make the following selections from the GIS Map window's menu bar:

Help ▶ Getting Started with SAS/GIS Software ▶ Begin Tutorial

The tutorial creates sample maps and attribute data and leads you step by step through the following tasks:

- displaying maps
- selecting the types of feedback that are provided about the displayed map
- using the zoom tool to zoom in on selected areas of the map
- using the pan tool to move the map within the window
- modifying, adding, and removing layers in the map
- using attribute data as a theme for a layer
- adding legends that explain how features are represented
- selecting features and using actions to explore the attribute data
- saving changes to the map
- geocoding addresses
- adding points to a map

After you have used the online tutorial to become familiar with the basics of using SAS/GIS software, you can refer to *SAS/GIS Software: Usage and Reference, Version 6* for additional information about using SAS/GIS software and for detailed reference information about the features of SAS/GIS software.

CHAPTER 2

Preparing Spatial Data

Assessing Your Spatial Data Needs

You use a geographic information system to explore data in the context of a map, so you must have a map in order to use SAS/GIS software. Furthermore, the map must be in the form of spatial data that SAS/GIS software can use.

Assessing Your Attribute Data

The first step in deciding what spatial data you need is to assess the attribute data that you want to analyze. The attribute data must have a spatial component. That is, the data must contain at least one variable with values that relate to location. Examples include city, state, or country names or codes; street names; addresses; and so on. Since SAS/GIS software is part of SAS, the attribute data must also be in the form of a SAS data set or a SAS view. If needed, you can use any method that is available for transforming your attribute data into a SAS data set or a SAS view. These methods include, but are not limited to the following:

- using SAS programming statements or the SAS Import Wizard to read external files into SAS data sets
- using SAS/ACCESS software or the SQL procedure to create views to database files
- using SAS programming statements or the SQL procedure to create dynamic views to SAS data sets.

After you have ensured that your attribute data has a spatial component and is in a format that SAS/GIS can read, you can move on to identifying and locating your spatial data.

Determining Your Spatial Data Requirements

In order to analyze attribute data with SAS/GIS software, you need spatial data that contains representations of features to at least the same level of detail as the location information in your attribute data. For example, if your attribute data consists of demographic data for states, then your spatial data must provide at least state boundaries. If your attribute data consists of demographic data for smaller census tracts, then you need spatial data that contains the corresponding census tract boundaries in order to explore the demographic data with SAS/GIS software.

Locating a Source of Spatial Data

You might be able to purchase appropriate spatial data that has already been prepared in SAS/GIS format by a commercial data vendor. Contact SAS Technical Support for information on sources for spatial data in SAS/GIS format.

The other way to acquire spatial data for use with SAS/GIS software is to import it from other formats. One readily accessible source of maps for importing is the map data sets that are provided with SAS/GRAPH software. However, these maps provide only political boundaries and not other physical features such as rivers and major highways. Other sources for spatial data that you can import include the following:

- Governmental agencies. For example, SAS/GIS software can import spatial data from TIGER/Line files produced by the U.S. Census Bureau* and from DLG files produced by the U.S. Geological Survey.
- Drawing and computer-aided design (CAD) packages. SAS/GIS software can import the DXF interchange format that is supported by products from various vendors.
- Tele Atlas N.V. SAS/GIS can import the Dynamap files.
- MapInfo Corporation. SAS/GIS can import MapInfo MIF and MID files.
- ArcInfo software by ESRI. SAS/GIS can import uncompressed ArcInfo interchange (E00) files.
- User-created files. If no other source is available, you can use SAS programming statements to convert your spatial data into the required generic format, which SAS/GIS software can then import.

Whatever the source, the spatial data must have at least one variable with values that match values in the attribute data that you want to analyze. If necessary, you can use SAS to process either the attribute data or the spatial data. For example, if your attribute data contains state names and your spatial data contains state codes, you can use SAS programming statements to generate corresponding codes for the names. Likewise, if your attribute data and spatial data both have codes to identify areas in the map, but the two sets of data use different codes for the same areas, then you can use SAS programming statements to translate the coding schemes.

Examples of Common Spatial Data Tasks

The rest of this chapter contains examples of tasks common to preparing spatial data. Each of the examples builds upon the preceding examples. Use the DATA step

* SAS/GIS can import 2006 Second Edition TIGER/Line files from the U.S. Census Bureau and earlier releases of the TIGER/Line spatial data format. Contact SAS Technical Support for the latest information about the availability of support for importing 2007 TIGER/Line Shapefiles.

and data set information provided in “Importing Your Spatial Data” on page 15 to import a data set containing spatial data for the counties of North Carolina and South Carolina. Use this map to perform the actions described in the rest of the chapter.

Importing Your Spatial Data

Suppose you are given the task to determine the level of change in the county populations for the states of North Carolina and South Carolina. SAS/GIS software provides you with the information that is collected in the MAPS.USAAC sample attribute data set. For each U.S. county, this data set has an observation that includes the following variables:

STATE
: The FIPS (Federal Information Processing Standards) code for the state. See “Using FIPS Codes and Province Codes” in *SAS/GRAPH: Reference* for more information on FIPS codes.

COUNTY
: The FIPS code for the county.

CHANGE
: The level of change in the county population.

In order to analyze the data in MAPS.USAAC, you need a map with corresponding state and county boundaries and compatible identifier values. The MAPS.COUNTY map data set that is supplied with SAS/GRAPH software has coordinates for U.S. state and county boundaries and also uses FIPS codes to identify states and counties.

To extract map data that contains only the required states, submit the following program in the Program Editor window:

```
data work.ncsc;
   set maps.county;
   where state in (37 45);   /* FIPS codes for NC and SC */
run;
```

To import the spatial data, open the GIS Spatial Data Importing window with the following selections from the GIS Map window’s menu bar: **File ▸ Import**

Specify the following information in the appropriate fields of the GIS Spatial Data Importing window.

`Import Type`	SASGRAPH
`SAS/GRAPH data set`	WORK.NCSC
`ID Vars`	STATE and COUNTY
Map Entries:	
`Library`	SASUSER
`Catalog`	NCSC
`Name`	NCSC
`Action`	Create
Spatial Data Sets:	
`Library`	SASUSER
`Name`	NCSC
`Action`	Create

The following display contains an example of the GIS Spatial Data Importing window with the information correctly entered in the fields.

Display 2.1 SAS/GIS Spatial Data Importing Window

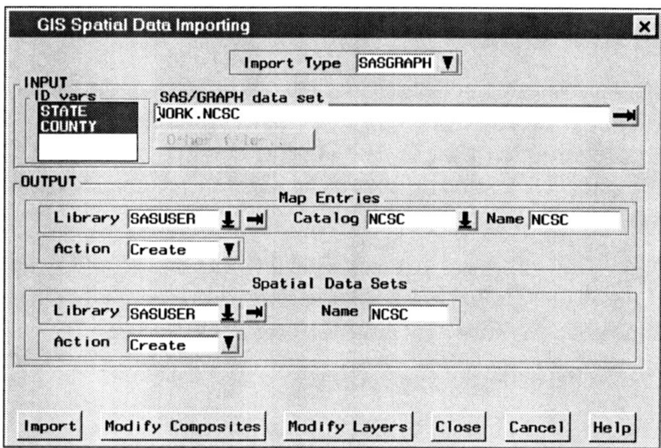

After entering these values, select **`Import`** to begin importing the spatial data.

When you receive the message **`Import complete. Close this window to display the map`** select **`Close`** to close the GIS Spatial Data Importing window. The imported map is now displayed in the GIS Map window, as shown in the following display.

Display 2.2 Initial Display of Imported SAS/GRAPH Map

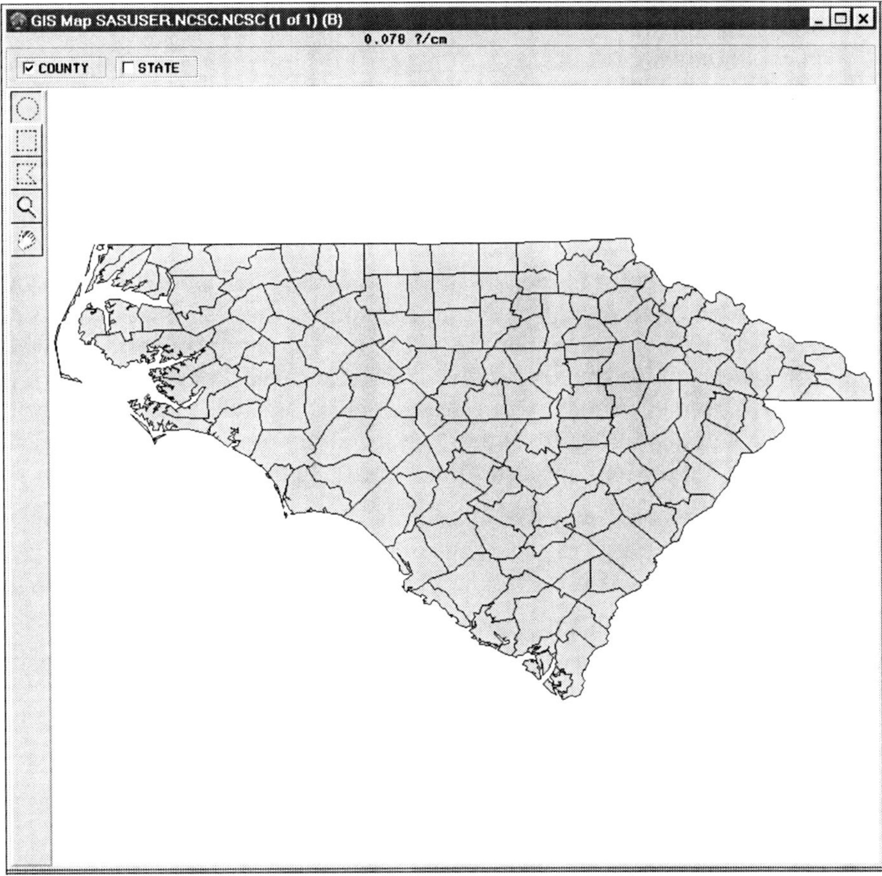

Note: For more information on importing spatial data from other formats, see Chapter 3, "Importing Spatial Data," on page 23. △

Changing the Default Characteristics of a Map

Once you have imported a new map or loaded a new SAS/GIS software spatial database from a commercial vendor, you might want to change some of the default characteristics of the map. Some of the characteristics you can change include the following:

- projection system used to display the map
- scale mode and units
- background color for the map area
- layers that are initially displayed or hidden
- level of detail for map features.

To learn about other ways in which you can customize the appearance of the map, see Chapter 5, "Customizing Maps," in *SAS/GIS Software: Usage and Reference, Version 6.*

Selecting a Map Projection

A projection is required to represent spherical features like the earth's surface on a flat medium like a display screen or printed page. SAS/GIS software supports a wide variety of projection methods, but it assumes by default that the coordinate values in newly imported spatial data are arbitrary Cartesian (X/Y) values (except for TIGER and DYNAMAP files, for which SAS/GIS software assumes latitude and longitude degrees). However, the coordinates in the MAPS.COUNTY map data set are actually latitude and longitude values in radians. As a result, the initial Carolinas map in Display 2.2 on page 17 is elongated and reversed right-to-left.

To change the projection system that is used for the spatial data, you use the GIS: Projection Options window. Open the GIS: Projection Options window with the following selections from the GIS Map window's menu bar: **Tools ▶ Map Properties ▶ Projections**

In the GIS: Projection Options window, `Storage Projection System` specifies the system that is used to interpret the stored spatial data, and `Display Projection System` specifies the system that is used to project the interpreted spatial data in the GIS Map window. Use the drop-down arrows to select `Lat/Lon` for both `Storage Projection System` and `Display Projection System`. Also in the storage system parameters, select `W` for `Hemisphere` and `1` for the `Units Multiplier`. Select `Close` to close the GIS: Projection Options window and apply the new projection specifications.

Note: This example uses the same projection system for the Storage Projection System and the Display Projection System. It is not required that the two use the same projection system unless the Storage Projection System is arbitrary Cartesian data. △

The spatial data is reloaded into the GIS Map window by using the new projection systems, and the projected version of the map is displayed as shown in the following display.

Display 2.3 Changing the Initial Projection

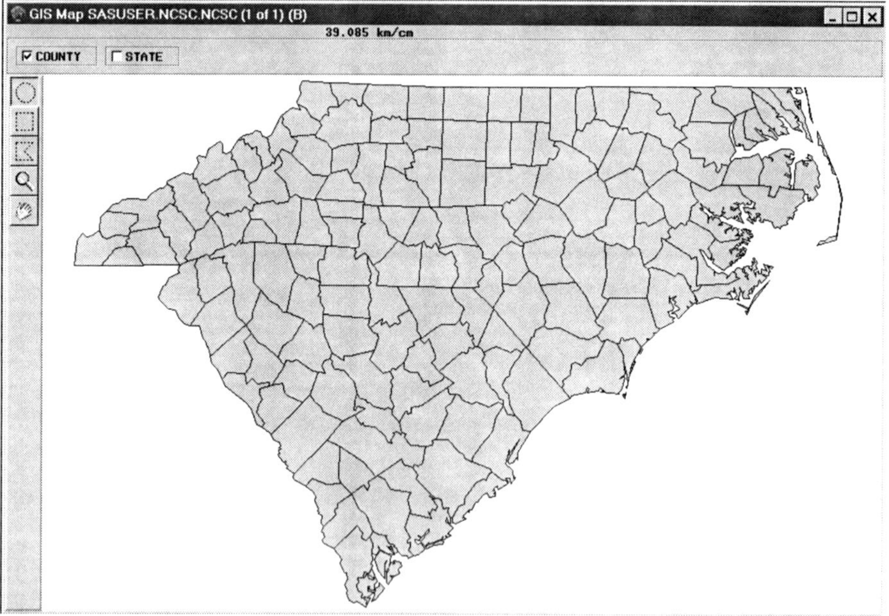

Selecting the Units System

By default, the scale feedback for a newly imported map uses metric units. You use the GIS Map Options window to set the units system of a new map. Open the GIS Map Options window with the following selections from the GIS Map window's menu bar: **Tools ▶ Map Properties ▶ Map Options**

Select **English** for the units system to change the scale mode to **mi/in** (miles per inch). Select **Close** to close the GIS Map Options window and apply the change to the map feedback area as shown in the following display.

Display 2.4 Changing the Initial Unit System

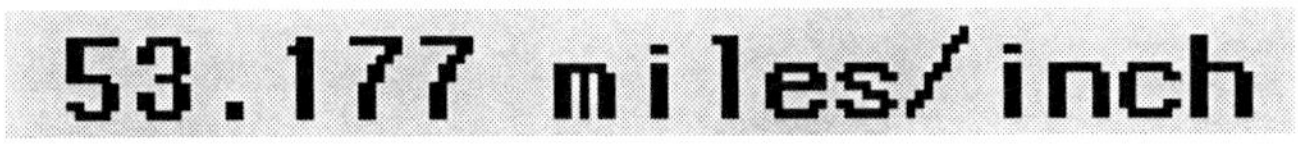

53.177 miles/inch

Selecting a Background Color

By default, a map area is assigned a white background. You use the GIS Map Styles and Colors window to choose a different background color for a map. Open the GIS Map Styles and Colors window with the following selections from the GIS Map window's menu bar:**Tools ▶ Map Properties ▶ Colors**

Use the drop-down arrow for **Background** to display a list of the standard SAS colors and select **Blue**. Select **Close** to close the window and apply the new color choice as shown in the following display.

Display 2.5 Changing the Initial Background Color

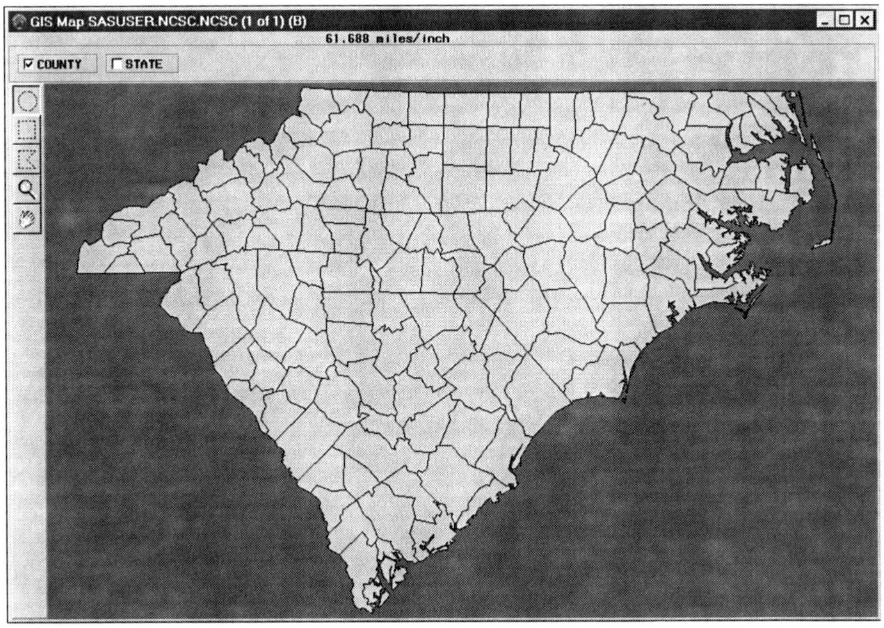

Choosing Which Layers Are Displayed

By default, only the first layer in the layer bar is displayed; other layers are hidden. To select which layers are displayed or hidden, click the corresponding layer bar check boxes. Deselecting a layer that is currently shown hides that layer, while selecting a layer that is currently hidden displays that layer.

In the example map, select the appropriate check boxes to hide the COUNTY layer and display the STATE layer as shown in the following display. When you are finished viewing the STATE layer, turn the COUNTY layer back on.

Display 2.6 Changing the Initial Active Layers

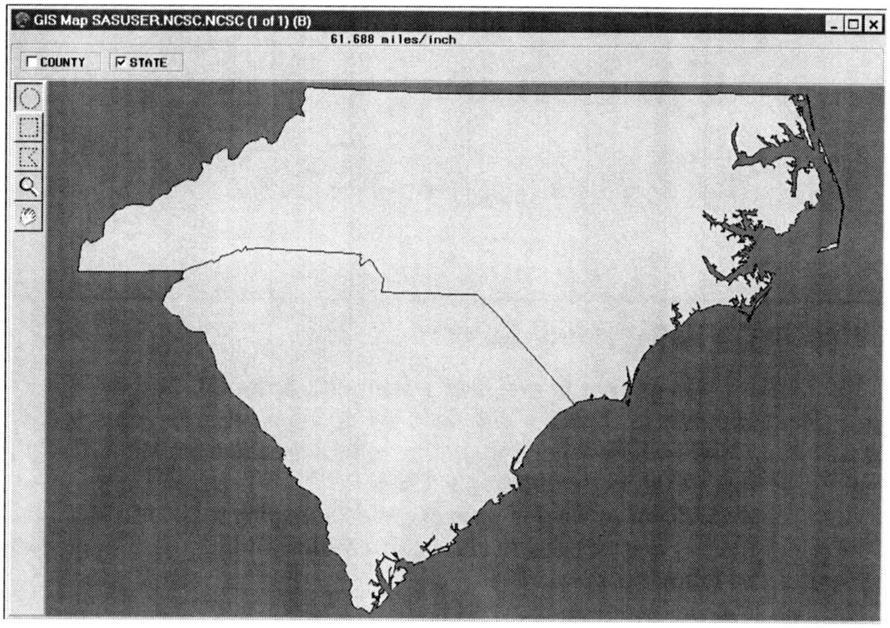

GIS Map SASUSER.NCSC.NCSC (1 of 1) (B)
61.688 miles/inch
COUNTY
STATE

Changing the Level of Detail

In spatial databases, SAS/GIS software distinguishes between the coordinate points that are necessary to represent features minimally and those that provide extra detail. For example, the starting and ending intersections of a segment of a street are considered fundamental points, while additional points that represent the curves between the intersections are considered extra detail. By default, SAS/GIS software uses detail points for all layers if they are available. To turn off the detail points for all features in the map, make the following selections from the GIS Map window menu bar:**View ▶ Detail**

With the detail turned off, map features are drawn more coarsely but more quickly because fewer lines are drawn. The following display shows the resulting map. Turn the detail back on to provide full detail to the map.

Display 2.7 Changing the Initial Detail Level

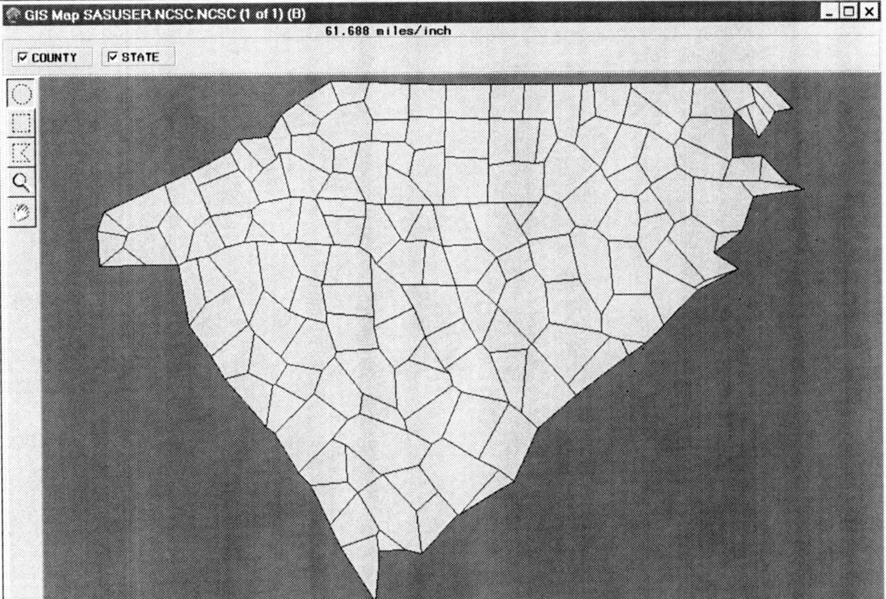

Linking the Attribute Data to the Spatial Data

Before you can use your spatial data as a basis for exploring your attribute data, you must link the attribute data to the spatial data. As explained in Chapter 1, "Overview of SAS/GIS Software," on page 3, one way to use the attribute data after you have linked it to the spatial data is by creating a *theme* to control the appearance of features in the spatial data.

In the layer bar, right-click the COUNTY layer name to open the pop-up menu for the COUNTY layer. Select `Edit` to open the GIS Layer window. In the definition for the COUNTY layer, select `Thematic`. The GIS Attribute Data Sets window opens for you to define the link to the theme data set.

In the GIS Attribute Data Sets window, select `New` to define a new link. In the resulting Select a Member window, select MAPS.USAAC. You must next specify the values that are common to both the attribute and spatial data, since the common values provide the connection between the spatial data and the attribute data. The spatial database and the MAPS.USAAC data set share compatible state and county codes, so first select STATE in both the `Data Set Vars` and `Composites` lists, and then select COUNTY in both lists. Select `Save` to save the link definition to the `Links` list. Finally, select `Continue` to close the GIS Attribute Data Sets window.

After the GIS Attribute Data Sets window closes, the Var window automatically opens for you. Select which variable in the attribute data provides the theme data for your theme. Select the CHANGE variable to have the counties colored according to the level of change in the county population. Select `OK` to close the Var window.

The counties in the spatial data are colored according to the demographic values in the attribute data set, as shown in the following display.

Display 2.8 Linking the Attribute Data as a Theme

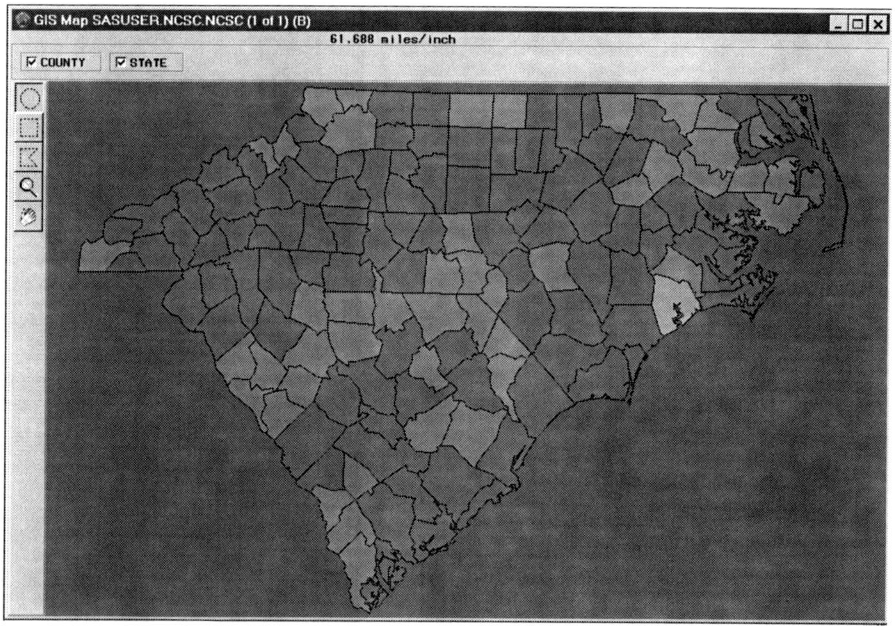

Note: The theme ranges in the COUNTY layer reflect the range of values in the MAPS.USAAC data set, which contains data for the entire United States. See Chapter 5, "Customizing Maps," in *SAS/GIS Software: Usage and Reference, Version 6* for details on how you can select different theme ranges that are more appropriate for the displayed counties. △

Saving the Map Characteristics

Changes that you make while the map is displayed are not automatically stored in the spatial database. To record these modifications for use in future sessions, you must write them to the spatial database. You can save all changes by making the following selections from the GIS Map window's menu bar:**File ▶ Save ▶ All**

As a safeguard, SAS/GIS software also offers you the choice of saving changes when you attempt to close the map.

CHAPTER

3

Importing Spatial Data

Overview of Importing Spatial Data

SAS/GIS software organizes spatial databases into SAS data sets and SAS catalog entries. Spatial data might be available from some vendors in the required SAS/GIS format, but any spatial data that is not in this format must be imported before it can be used with SAS/GIS software. SAS/GIS software provides interactive facilities for importing spatial data from the following formats:

uncompressed ArcInfo interchange files (E00)
produced by ArcInfo software from ESRI.

Digital Line Graph files (DLG)
from the U.S. Geological Survey and commercial data vendors.

Drawing Interchange Files (DXF)
produced by a variety of mapping and CAD software applications.

Dynamap files
from Tele Atlas N.V.

SAS/GRAPH map data sets
provided with SAS/GRAPH software.

Topologically Integrated Geographic Encoding and Referencing files (TIGER)
from the U.S. Census Bureau and commercial data vendors.

MapInfo files (MIF and MID)
from MapInfo Corporation.

SAS/GIS software also supports a generic format to accommodate other sources of spatial data for which no explicit importing facility is provided. You can use SAS programming statements to translate your spatial data into the generic format and then use SAS/GIS software to complete the process of importing it into a SAS/GIS spatial database. See "Importing Generic Spatial Data" on page 34 for more information on the generic import types.

SAS/GIS provides both interactive and programmatic ways to import spatial data. The remainder of this chapter explains how to import spatial data interactively using the GIS Spatial Data Importing Window. For information about how to import spatial data programmatically, see Chapter 4, "Batch Importing," on page 43.

The GIS Spatial Data Importing Window

The GIS Spatial Data Importing window provides an interactive facility for importing spatial data from other formats into SAS/GIS spatial databases. You use the GIS Spatial Data Importing window to specify the type of spatial data to import. To open the GIS Spatial Data Importing window, select **File ▶ Import** from the GIS Map window's menu bar, or select `Import` from the map pop-up menu when no map is displayed.

Display 3.1 GIS Spatial Data Importing Window

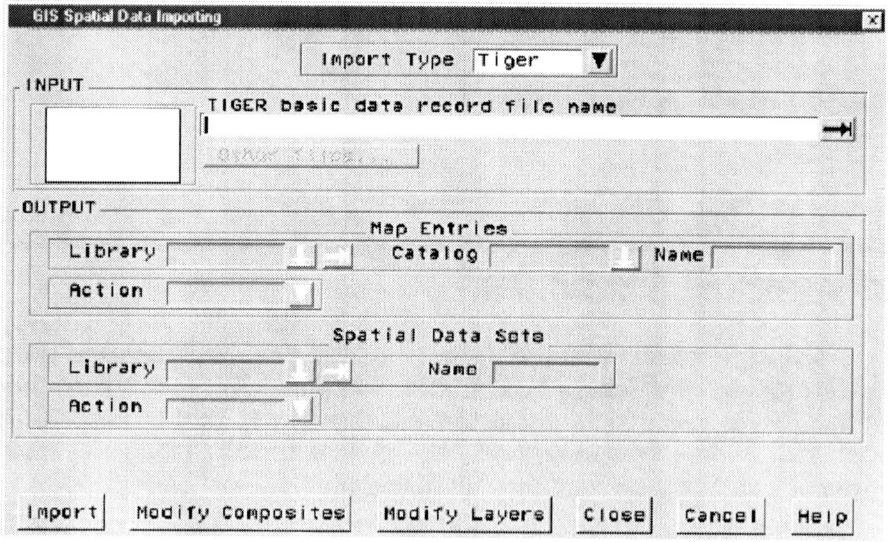

Elements of the Importing Window

The GIS Spatial Data Importing window is divided into the following areas:

- Import Type

□ Input
□ Output
□ command buttons

The Import Type Area

You use the **Import Type** field to specify the type of data you want to import. Click the arrow to the right of the **Import Type** field to display a list of available data types. Select a data type from this list to display it in the field.

The Input Area

You use the Input area of the window to specify the location of the spatial data files that you want to import. Additional information about each type of data is presented later in this chapter.

The Input area contains the following three elements:

□ A list, which can contain the names of variables you can select as ID variables or LAYER variables. ID variables apply only to SASGRAPH and GENPOLY import types. LAYER variables apply only to GENLINE and GENPOINT import types.

□ A filename field, which displays the name of the spatial data file. You can either type the name in the field, or click the arrow to display the Open window and then select the file.

  Note: If you are importing an external file, the arrow opens the Open window. However, if the import type is SASGRAPH or generic (SAS data set), the Select a Member window appears so that you can select a SAS data set. △

□ An **Other Files** button, which allows you to select other files that are associated with the main spatial data file.

  Note: The **Other Files** button usually applies only to the TIGER, MAPINFO, and DYNAMAP import types. Other data types might not have any additional spatial data files. △

The Output Area

The Output area of the window contains the following two sections:

Map Entries

You specify the storage location of the spatial database in this area.

In the **Library** field, you specify a name for the library that you want to contain the catalog and its entries. You can type the name of an existing library in the field, use the drop-down menu to select an existing libref, or use the pull-out menu to assign a new libref.

In the **Catalog** field, you specify a name for the SAS catalog that you want to contain the spatial database entries. You can type the name of a new catalog to be created, type the name of an existing catalog, or use the drop-down menu to select an existing catalog.

In the **Name** field, you specify a name for the GIS map. By default, the **Name** field contains the name of the spatial data import type, for example, TIGER, ARC, DXF, and so on. You can supply your own GIS map name. This name is also used for the coverage entry and as the base name of the polygonal index data set by using the first three letters of the name, and then by adding up to the first five letters of the layer name.

In the **Action** field, you select options regarding the catalog entries. The drop-down menu to the right of the field contains the following options:

CREATE
: creates new catalog entries.

REPLACE
: overwrites existing catalog entries. REPLACE will create new catalog entries if no entries exist for it to overwrite.

UPDATE
: updates existing catalog entries.

Spatial Data Sets
: In this area, you specify the name of the SAS library in which the chains, nodes, details, and polygonal index data sets are stored along with the base name for the chains, nodes, and details data sets and spatial entry.

In the **Library** field, you specify a name for the library that you want to contain the data sets. You can type the name of an existing library in the field, use the drop-down menu to select an existing libref, or use the pull-out menu to assign a new libref.

In the **Name** field, you specify a base name for the data sets. The chains, nodes, and details data set names are formed by adding a C, N, or D, respectively, to this base name. The base name is also used as the name for the spatial entry in the catalog.

In the **Action** field, you select options regarding the data sets. The drop-down menu to the right of the field contains the following options:

CREATE
: creates new data sets.

REPLACE
: overwrites existing data sets. REPLACE creates new data sets if no data sets exist for it to overwrite.

APPEND
: appends to existing data sets.

Command Buttons

The command buttons appear in a row along the lower edge of the GIS Spatial Data Importing window. You use these buttons to perform actions on your data sets, open the SAS/GIS online help, or close the window. The following list describes the different functions of the command buttons:

Import
: starts the importing process, provided that all required information has been supplied.

Modify Composites
: opens a window to view and modify the default composites that are created during the import.

Modify Layers
: opens a window to view and modify the default layer definitions that are created during the import.

Close
: closes the GIS Spatial Data Importing window and returns to the GIS Map window with the imported map displayed.

`Cancel`
closes the GIS Spatial Data Importing window. If a map was imported, it is not displayed in the GIS Map window.

`Help`
opens the online Help facility for the GIS Spatial Data Importing window. The Help provides details on the steps for importing the corresponding type of spatial data.

Common Importing Procedures

The following instructions detail the process that is common to importing all accepted types of spatial data. For additional information about preparing and importing specific types of data, see the sections that follow.

To import spatial data, complete the following steps:

1 Select **File ▶ Import** from the GIS Map window's menu bar or select `Import` from the map pop-up menu when no map is displayed.

The GIS Spatial Data Importing window appears.

2 Select the type of data to import from the `Import Type` field drop-down menu.

3 Select the file to import from the pull-out menu to the right of the filename field in the Input area. Or, supply the path and filename or SAS data set name in this field.

After you have selected an input file, the Output fields are filled with the default values. See "The SASHELP.GISIMP Data Set" on page 40 for information about changing the default values.

4 Modify the default composites, if needed. Click `Modify Composites` and make your changes in the Import window that appears. This step is optional. See "Defining Composites in Imported Data" on page 36 for more information.

Note: The composite definitions in the Import window can have a status of either Keep or Drop. Associations with the Drop status are not included in the SAS/GIS spatial database even though they appear in the variable list in the GIS Import window. △

5 Modify the layers, if needed. Click `Modify Layers` and make your changes in the Import window that appears. This step is optional. See "Defining Layers in Imported Data" on page 38 for more information.

6 Modify the `Library`, `Catalog`, `Name`, and `Action` field information in the Map Entries area to specify the destination location of the catalog and its entries, if needed.

Note: The `Library`, `Catalog`, `Name`, and `Action` fields contain default values that are based on the input file that you selected earlier in this process. You can modify these values or accept the defaults. △

7 Modify the `Library`, `Name`, and `Action` field information in the Spatial Data Sets area to specify the destination location of the spatial data sets, if needed.

Note: These fields contain default values that are based on the input file you selected earlier in the import process. You can modify these values or accept the defaults. △

8 Click the `Import` button.

Once the import has finished, the following message appears in the window message bar:

```
Import Complete.  Close this window to display the map.
```

CAUTION:

Be careful when you are using **Modify Composites** and **Modify Layers**, especially when appending new information to an existing map. Modifying the default composites and layers can cause unexpected results or errors. If you are unsure about making modifications, you should keep the default composites and layers as they are during importing. After you import the map, you can view it and review the composites and layers that were created and then use PROC GIS to make modifications later. △

Importing ArcInfo Interchange Data

ArcInfo software supports several spatial data formats, but SAS/GIS software can import only spatial data that has been exported from ArcInfo software in uncompressed interchange format (sometimes called E00 files because the files have the extension **.e00** by default). If you do not know whether a file is compressed, open the file with a host editor. If you can read text in the file, it is not compressed.

To import spatial data in uncompressed ArcInfo interchange format using the GIS Spatial Data Importing window, complete the following steps:

1 Select **ARC** from the **Import Type** drop-down menu.

   The title of the filename field in the Input area changes to **ARC/INFO Coverage export filename**.

2 Specify the path to the desired ArcInfo file, either by typing the path in the filename field, or by clicking the arrow to display an Open window and then selecting the file from that window.

   By default, SAS/GIS expects ArcInfo interchange files to have an extension of **.e00**.

   SAS/GIS allocates the SAS fileref ARCIN to the import path that you specified in the filename field. If you allocate the fileref ARCIN to the desired file before you begin the import process, the filename and path appears in the filename field automatically.

3 Modify the destination information for the catalog and the spatial data sets, if needed.

4 Modify the default layers and composites, if needed. See "Defining Composites in Imported Data" on page 36 and "Defining Layers in Imported Data" on page 38 for more information.

5 Click **Import** to import the data. When the import process is complete, a message is displayed in the window message bar to indicate whether the import was successful. You can proceed with another import or close the window to display the newly imported map.

Importing DLG Data

To import spatial data in Digital Line Graph (DLG) format using the GIS Spatial Data Importing window, complete the following steps:

1 Select **DLG** from the **Import Type** drop-down menu.

The title of the filename field in the Input area changes to **Digital Line Graph filename**.

2 Specify the path to the desired DLG file, either by typing the path in the filename field, or by clicking the arrow to display an Open window and then selecting the file from that window. SAS/GIS software checks whether the DLG file type is Standard or Optional and processes the types accordingly.

SAS/GIS allocates the SAS fileref DLGIN to the import path that you specified in the filename field. If you allocate the fileref DLGIN to the desired file before beginning the import process, the filename and path appear in the filename field automatically.

3 Modify the destination information for the catalog and the spatial data sets, if needed.

4 Modify the default layers and composites, if needed. See "Defining Composites in Imported Data" on page 36 and "Defining Layers in Imported Data" on page 38 for more information. No more than one layer of each type can be created from a DLG file. The fields for any layer types that cannot be created are dimmed.

5 Click **Import** to import the data. When the import process is complete, a message is displayed in the window message bar to indicate whether the import was successful. You can proceed with another import or close the window to display the newly imported map.

Importing DXF Data

Drawing Interchange File (DXF) files are typically output from CAD systems. DXF files often contain only lines and points. If you want to create polygons in the imported SAS/GIS map, then you must ensure that the boundary lines for the closed areas in the DXF file are topologically correct. Although the lines might appear to form a closed polygon in the CAD system, the polygon creation process will fail if the end point of one boundary line is not the same as the beginning point of the next line.

The SAS/GIS import process does not support DXF symbols or blocks. If parts of the imported drawing do not appear as expected, then examine the source of the DXF file. If it contains AutoCAD blocks, then the data provider can explode these blocks into separate elements and then export a new DXF file.

To import spatial data in DXF format using the GIS Spatial Data Importing window, complete the following steps:

1 Select **DXF** from the **Import Type** drop-down menu.

The title of the filename field in the Input area changes to **DXF filename**.

2 Specify the path to the desired DXF file, either by typing the path in the filename field, or by clicking the arrow to display an Open window and then selecting the file from that window.

SAS/GIS allocates the SAS fileref DXFIN to the import path that you specified in the filename field. If you allocate the fileref DXFIN to the desired file before beginning the import process, the filename and path appear in the filename field automatically.

3 Modify the destination information for the catalog and the spatial data sets, if needed.

4 Modify the default layers and composites, if needed. See "Defining Composites in Imported Data" on page 36 and "Defining Layers in Imported Data" on page 38 for more information.

5 Click **Import** to import the data. When the import process is complete, a message is displayed in the window message bar to indicate whether the import was successful. You can proceed with another import or close the window to display the newly imported map.

Importing Dynamap Data

To import spatial data in Dynamap format using the GIS Spatial Data Importing window, complete the following steps:

1 Select **Dynamap** from the **Import Type** drop-down menu.

The title of the filename field in the Input area changes to **Dynamap basic data record filename**.

2 Specify the path to the desired Dynamap basic data record file, either by typing the path in the filename field, or by clicking the arrow to display an Open window and then selecting the file from that window. The basic data record file is a Type 1 Dynamap file, and it contains a record for each line segment in the file. This file is required.

SAS/GIS software allocates the SAS fileref GDT1 to the import path that you specified in the filename field. If you allocate the fileref GDT1 to the desired file before beginning the import process, the filename and path appear in the filename field automatically.

3 Click **Other Files**

A window appears and displays filename fields for the remainder of the Dynamap files that are needed. If the files are in the same directory as the basic data record file, the path is specified automatically. Specify the following files:

Shape coordinate points
: provide additional coordinates that describe the shape of each line segment, for example, a curve in the road. These coordinates are SAS/GIS detail points. This file is a Type 2 Dynamap file and has a corresponding fileref of GDT2. This file is required.

Index to alternate feature names
: provides the names if a line segment has more than one feature name, for example, Main St. and State Highway 1010. This file is a Type 4 Dynamap file and has a corresponding fileref of GDT4. This file is optional and is not selected by default. To read in the data from this file, select the **Read On/Off** check box.

Feature name list
: provides a list of all unique feature names. This file is a Type 5 Dynamap file and has a corresponding fileref of GDT5. This file is optional and is not selected by default. To read in the data from this file, select the **Read On/Off** check box.

Additional address and ZIP code data
: provides additional address range information if the address information cannot be presented as a single address range. This file is a Type 6 Dynamap

file and has a corresponding fileref of GDT6. This file is optional and is not selected by default. To read in the data from this file, select the **Read On/Off** check box.

Click **OK** when you have specified the paths for the Dynamap files to return to the GIS Spatial Data Importing window.

4 Modify the destination information for the catalog and the spatial data sets, if needed.

5 Modify the default layers and composites, if needed. See "Defining Composites in Imported Data" on page 36 and "Defining Layers in Imported Data" on page 38 for more information.

6 Click **Import** to import the data. When the import process is complete, a message is displayed in the window message bar to indicate whether the import was successful. You can proceed with another import or close the window to display the newly imported map.

Importing MapInfo Data

To import spatial data in MapInfo (MIF and MID) format using the GIS Spatial Data Importing window, complete the following steps:

1 Select **Mapinfo** from the **Import Type** drop-down menu.

The title of the filename field in the Input area changes to **MAPINFO MIF filename**.

2 Specify the path to the desired MapInfo MIF file, either by typing the path in the filename field, or by clicking the arrow to display an Open window, and then selecting the file from that window. The MIF file has an extension of `.mif` and contains graphic objects.

SAS/GIS allocates the SAS fileref MIF to the import path that you specified in the filename field. If you allocate the fileref MIF to the desired file before beginning the import process, the filename and path appear in the filename field automatically.

3 Click **Other Files**, and then enter or select the path for the MapInfo MID file in the window that appears. If the MID file is in the same directory as the MIF file, SAS/GIS software automatically sets the path to the MID file. The MID file has an extension of `.mid` and contains tabular data. SAS/GIS allocates the SAS fileref MID to the import path that you specified in the filename field. If you allocate the fileref MID to the desired file before beginning the import process, the filename and path appears in the filename field automatically.

Click **OK** when you have specified the path for the MID file to return to the GIS Spatial Data Importing window.

4 Modify the destination information for the catalog and the spatial data sets, if needed.

5 Modify the default layers and composites, if needed. See "Defining Composites in Imported Data" on page 36 and "Defining Layers in Imported Data" on page 38 for more information.

6 Click **Import** to import the data. When the import process is complete, a message is displayed in the window message bar to indicate whether the import was successful. You can proceed with another import or close the window to display the newly imported map.

Importing SAS/GRAPH Map Data Sets

To import spatial data in SASGRAPH format using the GIS Spatial Data Importing window, complete the following steps:

1 Select **SASGRAPH** from the **Import Type** drop-down menu.

  The title of the filename field in the Input area changes to **SAS/GRAPH data set**.

2 Specify the library and data set name of the desired SAS/GRAPH data, either by typing the data set name in the data set name field, or by clicking the arrow to display a Select A Member window and then selecting the data set from that window.

3 Select the variables from the **ID Vars** field that you want to use as ID variables. ID variables are variables whose values uniquely identify unit areas in the map. Typical ID variables in SAS/GRAPH maps are COUNTRY, ID, STATE, and COUNTY. A separate layer is created for each ID variable. The ID variables must be selected in hierarchical order. For example, if the data set contains both STATE and COUNTY variables, then STATE must be selected before COUNTY.

4 Modify the destination information for the catalog and the spatial data sets, if needed.

5 Modify the default layers and composites, if needed. See "Defining Composites in Imported Data" on page 36 and "Defining Layers in Imported Data" on page 38 for more information.

6 Click **Import** to import the data. When the import process is complete, a message is displayed in the window message bar to indicate whether the import was successful. You can proceed with another import or close the window to display the newly imported map.

Importing TIGER Data

To import spatial data in the 2006 Second Edition of Topologically Integrated Geographic Encoding and Referencing (TIGER) format using the GIS Spatial Data Importing window, complete the following steps:

1 Select **Tiger** from the **Import Type** drop-down menu.

  The title of the filename field in the Input area changes to **TIGER basic data record filename**.

2 Specify the path to the desired TIGER basic data record file, either by typing the path in the filename field, or by clicking the arrow to display an Open window and then selecting the file from that window.

  Each TIGER map consists of a set of files with names of the form TGR*ssccc*.F*vn*, where

  ss — is the two-digit FIPS code for the state.

  ccc — is the three-digit FIPS code for the county.

  Note: Refer to the documentation that accompanies the TIGER data for a directory of the FIPS codes for each state and county. △

  v — identifies the TIGER version number.

n identifies the TIGER record type. The basic data record is Type 1.

SAS/GIS allocates the SAS fileref TIGER1 to the import path that you specified in the filename field. If you allocate the fileref TIGER1 to the desired file before beginning the import process, the filename and path appear in the filename field automatically.

3 Click **Other Files**, and then type or select the path for the other TIGER data files in the window that appears. If the files are in the same directory as the basic data record file, SAS/GIS software automatically sets the path to the other files. Specify the following files:

Shape coordinate points
: provides additional coordinates that describe the shape of each line segment, for example, a curve in the road. These coordinates are SAS/GIS detail points. This file is required and is selected by default. This file is a Type 2 TIGER file. SAS/GIS allocates the fileref TIGER2 to this path.

Index to alternate feature names
: provides the names if a line segment has more than one feature name, for example, Main St. and State Highway 1010. This file is optional and is not selected by default. To read in the data from this file, select the **Read On/Off** check box. This file is a Type 4 TIGER file. SAS/GIS allocates the fileref TIGER4 to this path.

Feature name list
: provides a list of all unique feature names. This file is optional and is not selected by default. To read in the data from this file, select the **Read On/Off** check box. This file is a Type 5 TIGER file. SAS/GIS allocates the fileref TIGER5 to this path.

Additional address and ZIP code data
: provides additional address range information if the address information cannot be presented as a single address range. This file is optional and is not selected by default. To read in the data from this file, select the **Read On/Off** check box. SAS/GIS allocates the fileref TIGER6 to this path.

When you have specified the paths for the TIGER files, click **OK**to return to the GIS Spatial Data Importing window.

4 Modify the destination information for the catalog and the spatial data sets, if needed.

5 Modify the default layers and composites, if needed. See "Defining Composites in Imported Data" on page 36 and "Defining Layers in Imported Data" on page 38 for more information.

Note: By default, the following composites are assigned Drop status and will not appear in the imported data*:

AIR	RECTYPE
ANC	SIDECODE
IADDR	SOURCE

* Refer to the documentation for TIGER/Line files for more information on these composites.

△

6 Click **`Import`** to import the data. When the import process is complete, a message is displayed in the window message bar to indicate whether the import was successful. You can proceed with another import or close the window to display the newly imported map.

Note: SAS/GIS can import 2006 Second Edition TIGER/Line files from the U.S. Census Bureau and earlier releases of the TIGER/Line spatial data format. Contact SAS Technical Support for the latest information about the availability of support for importing files from 2007 TIGER/Line Shapefiles. △

Importing Generic Spatial Data

SAS/GIS software provides facilities for creating spatial databases from SAS data sets that contains the following types of generic spatial data:

point (GENPOINT)
: consists of discrete points.

line (GENLINE)
: consists of discrete line segments.

polygon (GENPOLY)
: consists of areas that are enclosed by polylines.

You can use the generic import methods if your data is in a format other than the specific import types that were discussed earlier. The generic import methods are useful for combining map features with an existing map. However, when adding generic data to existing spatial data sets, you must ensure that coordinate systems match.

Importing Generic Point (GENPOINT) Data

To import a SAS data set that contains point data, use the GIS Spatial Data Importing window to complete the following steps:

1 Select **`Genpoint`** from the **`Import Type`** drop-down menu.

The title of the filename field in the Input area changes to **`SAS/GIS Generic Point data set`**.

2 Specify the desired SAS data set, either by typing the location in the data set field or by clicking the arrow to display the Select a Member window and then selecting the data set from that window.

The point data set must contain at least the following variables:

X
: east-west coordinate of the point.

Y
: north-south coordinate of the point.

ID
: identifier value for the point.

Note: Each observation in the data set must have a unique value for the ID variable. △

The data set can also contain other variables, for example, variables to define characteristics of the points.

3 Select the variable from the **`ID Vars`** field that you want to use as an ID variable. SAS/GIS software performs a frequency analysis on the values of the specified

variable in the point data set and creates a point layer for each unique value of the specified variable. If you specify more than 16 layers, only the first 16 are added to the map. If you do not specify a layer variable, the resulting map will have a single point layer with the same name as the original point data set.

4 Modify the destination information for the catalog and the spatial data sets, if needed.

5 Modify the default layers and composites, if needed. See "Defining Composites in Imported Data" on page 36 and "Defining Layers in Imported Data" on page 38 for more information.

6 Click **Import** to import the data. When the import process is complete, a message is displayed in the window message bar to indicate whether the import was successful. You can proceed with another import or close the window to display the newly imported map.

Importing Generic Line (GENLINE) Data

To import a SAS data set that contains line data, use the GIS Spatial Data Importing window to complete the following steps:

1 Select **Genline** from the **Import Type** drop-down menu.

The title of the filename field in the Input area changes to **SAS/GIS Generic Line data set**.

2 Specify the desired SAS data set, either by typing the location in the data set field, or by clicking the arrow to display the Select a Member window and then selecting the data set from that window.

The line data set must contain at least the following variables:

X
: east-west coordinate of a point on the line.

Y
: north-south coordinate of a point on the line.

ID
: identifier value for the line.

Note: Each line in the data set must have a unique ID value, and all observations for the points on each line must have the same value for the ID variable. △

The data set can also contain other variables, for example, variables to define characteristics of the lines.

3 Select the variable from the **ID Vars** field that you want to use as an ID variable. SAS/GIS software performs a frequency analysis on the values of the specified variable and creates a line layer for each unique value of the specified variable. If more than 16 layers are created, only the first 16 are added to the map by default. If you do not specify a layer variable, the resulting map will have a single line layer with the same name as the original line data set.

4 Modify the destination information for the catalog and the spatial data sets, if needed.

5 Modify the default layers and composites, if needed. See "Defining Composites in Imported Data" on page 36 and "Defining Layers in Imported Data" on page 38 for more information.

6 Click **Import** to import the data. When the import process is complete, a message is displayed in the window message bar to indicate whether the import was successful. You can proceed with another import or close the window to display the newly imported map.

Importing Generic Polygon (GENPOLY) Data

To import a SAS data set that contains polygon data, use the GIS Spatial Data Importing window to complete the following steps:

1 Select `Genpoly` from the `Import Type` drop-down menu.

The title of the filename field in the Input area changes to `SAS/GIS Generic Polygon data set`.

2 Specify the desired SAS data set, either by typing the location in the data set field, or by clicking the arrow to display the Select a Member window and then selecting the data set from that window.

The polygon data set must contain at least the following variables:

X
: east-west coordinate of a point on the polygon boundary.

Y
: north-south coordinate of a point on the polygon boundary.

ID-name(s)
: identifier value(s) for the polygonal area.

Note: Each polygonal area in the data set should have unique identifier values, and all observations for the points in each area should have the same identifier value. A polygonal area can consist of more than one polygon. In that case the data set should also contain a SEGMENT variable to distinguish the individual polygons. △

Any other variables in the data set will not be included in the spatial database.

3 Select the variables from the `ID Vars` field that you want to use as ID variables. ID variables are variables whose values uniquely identify unit areas in the map. A separate layer is created for each ID variable. The ID variables must be selected in hierarchical order. For example, if the data set contains both STATE and COUNTY variables, then STATE must be selected before COUNTY.

4 Modify the destination information for the catalog and the spatial data sets, if needed.

5 Modify the default layers and composites, if needed. See "Defining Composites in Imported Data" on page 36 and "Defining Layers in Imported Data" on page 38 for more information.

6 Click `Import` to import the data. When the import process is complete, a message is displayed in the window message bar to indicate whether the import was successful. You can proceed with another import or close the window to display the newly imported map.

Defining Composites in Imported Data

As a preliminary step to actually importing your data, the import process identifies all composites that will be created by the import. A composite defines the role that a variable (or variables) plays in the spatial data, and how it should be used to represent features on the map. With the exception of the TIGER and DYNAMAP import types, which have a standard set of predefined composites, the composites are based on attributes that are found in the input data. The composites are actually created during the import. However, you have the opportunity to review the default composites before the import takes place, and you can modify them if you choose.

Once you have filled out the Input and Output sections on the GIS Spatial Data Importing window, you can click **`Modify Composites`** (before you click **`Import`**). This action will open the Import window as shown in the following display.

Display 3.2 The Import Window for Defining Composites

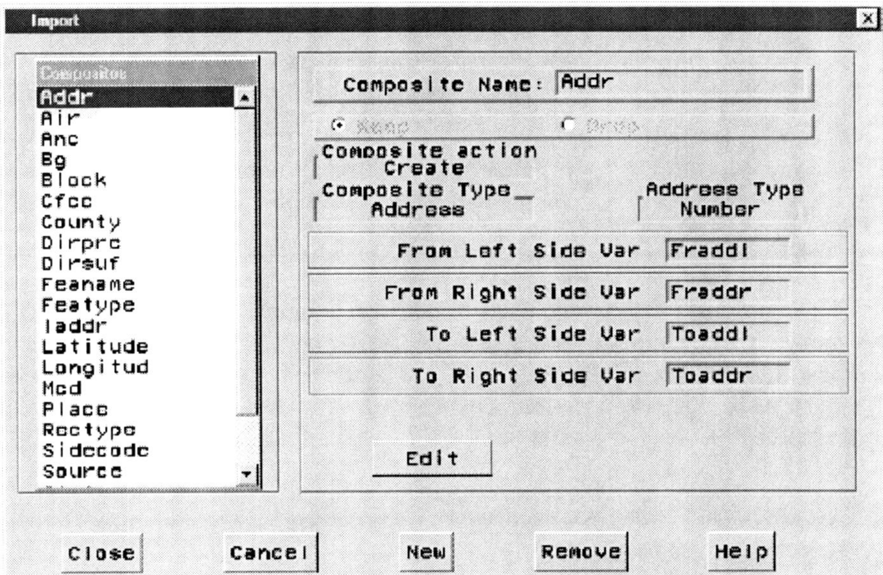

When the window is first displayed, only the Composites list is shown. This list contains all of the composites that will be created for the import, in addition to composites that are found in maps that are being appended to. To review a composite, click its name in the Composites list. The other fields in the window are then displayed. The values in these fields are used to define the composite.

To modify the definition of the selected composite, click **`Edit`** and all the fields will become editable. When you are finished editing, click **`Save`** if you want to save your changes or **`Cancel`** if you do not want to save the changes.

The following table describes each of the fields and their uses.

Table 3.1 Import Window Elements for Modifying Composites

Window Elements	Use
`Composite Name`	This field allows you to change the composite name.
`Keep/Drop`	If Drop is selected, all the variables that define the composite are dropped from the spatial data sets; otherwise, they are kept.
`Composite Action`	This is a noneditable field and notes if the composite will be created or replaced.
`Composite Type`	This field defines the composite type. Modify this value by clicking the arrow next to the field and selecting one of the types from the list.
`Address Type`	This field is visible only for the Address composite type. Modify this value by clicking the arrow next to the field and selecting one of the types from the list.

Window Elements	Use
SAS variables	The SAS variables that define the composite are listed in separate fields. There might be as many as four of these fields, depending on the composite type. You can either type in these fields, or click the arrow next to the field to access the list of available variables. You will notice, however, that the arrows are not visible until one of the other composites has been removed. Then, the list will contain the SAS variables that belonged to the composite that was removed.
Polygonal	This check box indicates whether the polygonal index data set will be created for this composite. This is valid only for Area type composites. Selecting this check box hides and displays the **Index DS** (Data Set) field.
Index DS	You can type in the name of the index data set or use the arrow to bring up the Select a Member window and then select a SAS data set.

The command buttons are used to perform window-wide functions. The **Close** button closes the window and saves all changes that you made. The **Cancel** button closes the window and cancels all changes that you made. The **New** button allows you to define a new composite. The **Remove** button removes the currently selected composite. The **Help** button accesses the Help system.

Defining Layers in Imported Data

In addition to identifying the default composites, a preliminary step of the import is to identify all of the layers that will be created by the import. Each layer represents a set of features on the map and how those features will be displayed. With the exception of the TIGER and DYNAMAP import types, which have a standard set of predefined layer definitions, the layer definitions are based on attributes that are found in the input data. The layers are actually created during the import. However, you can review the default layer definitions before the import takes place, and you can modify them if you choose.

Before clicking the **Import** button, you must fill out the Input and Output sections on the GIS Spatial Data Importing window. Then you can click the **Modify Layers** button, which opens the Import window as shown in the following display.

Display 3.3 The Import Window for Defining Layers

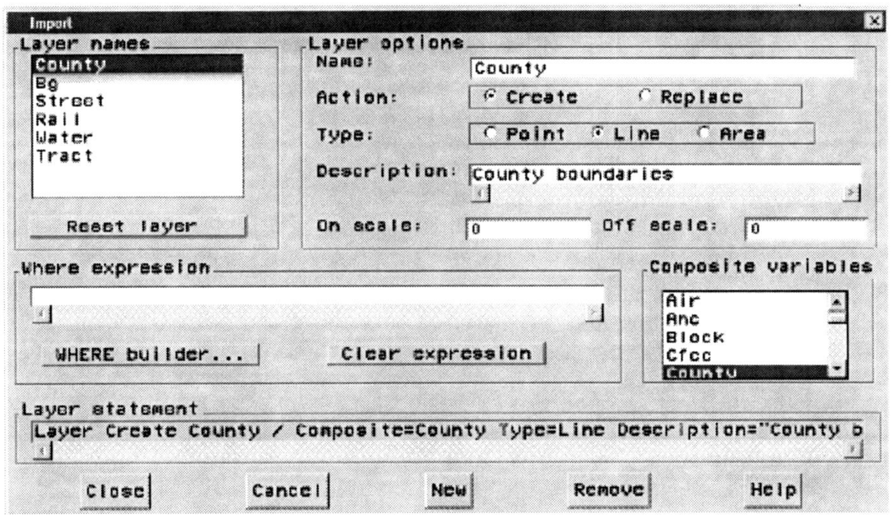

When the window is first displayed, the first layer in the Layer Names list is selected and its definition is displayed in the window. This list contains all of the layers that will be created for the import. To review a layer, select its name in the Layer Names list. All of the information in the window will be updated for the selected layer. You can now modify any of the fields. Unlike the Import window for modifying composites, all the fields in the window, except the **`Layer Statement`** field, are immediately editable, and none are hidden. The following table describes each of the fields and their uses.

Table 3.2 Import Window Elements for Modifying Layers

Window Elements	Use
Layer Names	Selectable list of layer names.
`Reset Layer`	Resets the layer definition to the way it was initially (with the exception of the layer name).
`Name`	To change the layer name, type another name in this field.
Action	Choices are either **`Create`** or **`Replace`**.
Type	Specifies the type of layer to be created or replaced. Choices are **`Point`**, **`Line`**, or **`Area`**. By default, one of these will be selected, but you can change it to either of the others.
`Description`	A description for the layer.
`On/Off Scale`	Defines the scale at which to turn the layer on and off. The default is 0.
`Where Expression`	Specifies the expression to be used to define features for the layer. You can type directly in this field. Invalid expressions are flagged.
`Where Builder`	Brings up the WHERE window, which allows you to build a valid WHERE expression. When the WHERE window is closed, the resulting expression is displayed in the **`Where Expression`** field.
`Clear Expression`	Clears the WHERE expression.

Window Elements	Use
Composite Variables	Contains all the composites that are defined for this import. Selecting a composite inserts it in the LAYER statement.
Layer Statement	A noneditable text field that displays the LAYER statement as it will appear when it is sent to the GIS procedure by the import.

The command buttons are used to perform window-wide functions. The **Close** button closes the window and saves all changes that you made. The **Cancel** button closes the window and cancels all changes that you made. The **New** button allows a new layer to be defined. The **Remove** button removes the currently selected layer. The **Help** button accesses the Help system.

After importing, you can use the GIS procedure to create additional composites and define new layers. For details, see Chapter 7, "The GIS Procedure," on page 85.

The SASHELP.GISIMP Data Set

You can manually change the values for the destination location each time you import data. However, you can also set new default values that will be in effect for each subsequent import. Use caution when changing these values because the import has predefined values in the SASHELP.GISIMP data set that are needed for the import to complete.

The SASHELP.GISIMP data set supplies values that you need to import your spatial data. Included in this data set are two variables, DEFMLIB and DEFSLIB, which are used to supply the default values for the Map Entries Library and the Spatial Data Sets Library. To specify different default values, complete the following steps:

1. Open the SASHELP.GISIMP data set (you must have WRITE access to the SASHELP library).
2. Change the values of the DEFMLIB and the DEFSLIB variables.
   a. Change the values of the DEFMLIB variable to the libref for the library that you want to use as the default for the map entries.
   b. Change the value of the DEFSLIB variable to the libref for the library that you want to use as the default for the spatial data sets.
3. Save the data set.
4. Open the GIS Spatial Data Importing window again.

If the specified library does not exist, an error message is issued, and the name SASUSER is substituted for the libref.

If you do not have WRITE access to the SASHELP library, and you want to change the default values for the DEFMLIB and DEFSLIB variables, then copy the SASHELP.GISIMP data set to an allocated library to which you do have WRITE access. Change the values and save the data set as described above. Before you open the GIS Spatial Data Importing window, you must assign the new location of the GISIMP data set to the macro variable **USER_FIL**. For example, if you copy the SASHELP.GISIMP data set to your SASUSER library, submit the following statement:

```
%LET USER_FIL=SASUSER.GISIMP;
```

If you want to use the default values for a particular import without having to modify the SASUSER.GISIMP data set, you can reset the USER_FIL macro variable to the default SASHELP.GISIMP data set. For example:

```
%LET USER_FIL=SASHELP.GISIMP;
```

The import will use the values in the data set that the USER_FIL macro variable points to.

CHAPTER 4

Batch Importing

Overview of Batch Importing

The SAS/GIS Batch Import process allows you to use SAS Component Language (SCL) code to import data into SAS/GIS without using the interactive GIS Spatial Data Importing window, or even invoking SAS/GIS. This feature can be useful when you have large amounts of data to import. For example, it lets you set up a batch job to run overnight.

The SAS/GIS Batch Import process allows you to define the values that are needed for the import through macro variables and SAS filerefs. After you define the values, you then call an SCL entry to actually initiate the import. The process has three main steps:

1. Specify the input parameters.

   Include the definitions of the type of data to import, the location of the input spatial data, and any other specifications for identification variables (not necessary for all import types). You define the input parameters either by setting the values of macro variables or by assigning filerefs (depending on the import type).

2. Specify the output parameters.

   Include the library in which the output spatial data sets and catalogs will be stored, name specifications for catalogs, data sets and catalog entries, and whether they will be created, replaced, or updated. You define all output parameters by setting the values of macro variables.

3. Initiate the batch import

   Execute the SASHELP.GISIMP.BATCH.SCL entry to start the Batch Import process. You do not pass any parameters directly to the SCL entry; the parameters must all have been defined through macro variables and filerefs before you call the SCL entry.

Implementation of the Batch Import Process

You specify the parameters for the import by assigning a value to a macro variable or by assigning a fileref, as indicated. There are several ways to assign a value to a macro variable, including the %LET statement, the SYMPUT SAS CALL routine in the DATA step, and the SYMPUT/SYMPUTN function in SCL. As long as the value you want is stored in the macro variable along with the required name, it does not matter which method you use. However, all examples in this section use the %LET statement.

There are also several ways to assign a fileref, including the FILENAME statement, the FILENAME function in SCL, and host-specific file allocation mechanisms. However, all examples in this section use a FILENAME statement.

Specifying the Input Parameters

The input parameters define the type of data to import, the location of the input spatial data, and other specifications for variables in your data set (not necessary for all import types).

The IMP_TYPE Macro Variable

You must define a macro variable named IMP_TYPE to indicate which type of data you are going to import. For example, to import a TIGER file, submit the following statement:

```
%let IMP_TYPE=TIGER;
```

This parameter is required. The following table contains valid values for IMP_TYPE.

Table 4.1 Valid IMP_TYPE Values and Descriptions

IMP_TYPE Value	Description
TIGER	Topologically Integrated Geographic Encoding and Referencing (TIGER) files from the U.S. Census Bureau and commercial data vendors (2006 Second Edition and earlier)
DYNAMAP	Dynamap files from Tele Atlas N.V.
DLG	Digital Line Graph (DLG) files from the U.S. Geological Survey and commercial data vendors
DXF	Drawing Interchange Files (DXF) produced by a variety of mapping and CAD software applications
ARC	Uncompressed ArcInfo interchange (E00) files from ESRI.
SASGRAPH	SAS/GRAPH map data set format
GENLINE	Generic Line format SAS data set
GENPOINT	Generic Point format SAS data set

IMP_TYPE Value	Description
GENPOLY	Generic Polygon format SAS data set
MAPINFO	MapInfo Interchange format files from MapInfo Corporation

The INFILE Macro Variable or Required Filerefs

You must specify where the input spatial data is located by using either the INFILE macro variable or the required filerefs for your import type.

For the Generic import types and the SASGRAPH import type, you accomplish this by assigning the name of a SAS data set to the INFILE macro variable. You can specify a one-level or a two-level name. One-level names are assumed to be located in the WORK library. For example:

```
/* The CUBA data set in the MAPS library. */
%let INFILE=MAPS.CUBA;
```

or

```
/* The NC data set in the WORK library. */
%let INFILE=NC;
```

For all other import types, you must allocate filerefs to point to the files that you want to import. See "File Reference Table for Batch Importing" on page 51 for additional information about the filerefs for each import type. The following table contains the import types and the filerefs that you are required to assign for them.

Table 4.2 Import Types and Their Corresponding Filerefs

Import Type	Fileref
TIGER	assigns the filerefs TIGER1 and TIGER2 for file types 1 and 2, respectively. File types 4, 5, and 6 are optional; these are allocated to the filerefs TIGER4, TIGER5, and TIGER6, respectively. See "Hints and Tips for Batch Importing" on page 52 for information about using the optional file types.
DYNAMAP	assigns the filerefs GDT1 and GDT2 for file types 1 and 2, respectively. File types 4, 5, and 6 are optional; these are allocated to the filerefs GDT4, GDT5, and GDT6, respectively. See "Hints and Tips for Batch Importing" on page 52 for information about using the optional file types.
DLG	assigns the fileref DLGIN to the Digital Line Graph (DLG) file that you want to import.
DXF	assigns the fileref DXFIN to the Drawing Interchange File (DXF) that you want to import.
ARC	assigns the fileref ARCIN to the uncompressed file that you want to import.
MAPINFO	assigns the filerefs MID and MIF to the MapInfo Interchange Format MID file and the MapInfo Interchange Format MIF file, respectively.

The NIDVARS and IDVAR*n* Macro Variables

Note: NIDVARS and IDVAR*n* macro variables are for SASGRAPH and Generic Import Types only. △

For SASGRAPH and all of the Generic import types, you must provide additional information about variables in your data set.

For SASGRAPH and the Generic Polygon import types, you must identify the number and names of the variables that uniquely identify unit areas in the map, in hierarchical order. The NIDVARS macro variable must be set to the number of identification variables, and the IDVAR*n* macro variables must specify, in order, the names of the identification variables. For example, for a SASGRAPH import of the MAPS.USCOUNTY data set (which contains State and County boundaries), you specify:

```
%let IMP_TYPE=SASGRAPH;
%let INFILE=MAPS.USCOUNTY;
%let NIDVARS=2;
%let IDVAR1=STATE;
%let IDVAR2=COUNTY;
```

The value of *n* in IDVAR*n* ranges from 1 to the value that is specified for the NIDVARS macro variable.

For the Generic Line and the Generic Point import types, you can identify a single variable that will be used to generate layer definitions. You set NIDVARS=1 and IDVAR1=*layer variable*. A layer is created for each unique value of the specified layer variable. If you specify more than 16 layers, only the first 16 layers are added to the map. If you want all features to be added to a single layer, specify NIDVARS=0. For example, to add all points from a data set to a single layer, specify

```
%let NIDVARS=0;
```

However, if your points represent stores and you have a SIZE variable that indicates whether the store is small, medium, or large, then you can specify the following to create three separate layers, one for each value of SIZE:

```
%let NIDVARS=1;
%let IDVAR1=SIZE;
```

The KEEPTEMP Macro Variable

During a batch import, numerous working data sets are created in the WORK library. By default, these intermediate data sets are deleted when the batch import is complete. If you encounter problems during a batch import, then you may want to retain these data sets to help identify the cause. To prevent the working data sets from being deleted, specify the following:

```
%let KEEPTEMP=1;
```

The AREA, CENTROID, and CENTROID_OPT Macro Variables

You can specify that the batch import calculate the surface area, perimeter, and centroid of the polygons when importing polygonal data. The following list contains descriptions of the AREA, CENTROID, and CENTROID_OPT macro variables:

AREA
: specifies whether to calculate the enclosed areas and perimeter lengths for the area composite. Setting the value to 0 indicates that the area and perimeter are not to be calculated. Setting the value to 1 indicates that the area and perimeter are to be calculated. The calculated area is added to the polygonal index data set in a variable named AREA. A label for the AREA variable contains the storage area units. The calculated perimeter is added to the polygonal index data set in a variable named PERIMETER. A label for the PERIMETER variable contains the units.

CENTROID
: specifies whether to calculate the centroid of a polygon. Setting the value to 0 indicates that centroids are not to be calculated. Setting the value to 1 indicates that centroids are to be calculated.

CENTROID_OPT
: specifies the type of centroid to be calculated when the CENTROID macro variable has a value of 1. Setting the value to GEOMETRIC requests the actual calculated centroid, which might fall within the boundaries of the corresponding polygon. Setting the value to VISUAL adjusts the centroid so that it falls within the boundaries of the corresponding polygon. The centroid coordinates are added to the polygonal index data set in variables that are named CTRX and CTRY. Labels for the CTRX and CTRY variables contain the storage projection units and indicate whether it is a geometric or visual centroid.

For example, to calculate the area and visual centroids, specify the following:

```
%let AREA=1;
%let CENTROID=1;
%let CENTROID_OPT=VISUAL;
```

Note: The area, perimeter, and centroids are not calculated unless these macro variables are defined. Once defined, these variables will be used by the Batch Import process until reset in the current SAS session. △

Specifying the Output Parameters

The output parameters define the locations where the output data sets and catalog entries are stored and whether they will be created, replaced, or updated. These parameters are required for all import types. The following table contains the macro variables and a description of the information each specifies.

Table 4.3 Output Parameters Macro Variables and Descriptions

Macro Variables	Description
MAPLIB	specifies the libref in which the catalog is stored.
MAPCAT	specifies the catalog in which the entries are stored.
MAPNAME	specifies the name for the map and coverage entries.
CATHOW	specifies the action to be used for the catalog entries. Valid values are CREATE, REPLACE, or UPDATE.
SPALIB	specifies the libref in which the spatial data sets are stored.

Macro Variables	Description
SPANAME	specifies the name of the spatial entry and the prefix for the spatial data sets.
SPAHOW	specifies the action to be used for the spatial data sets. Valid values are CREATE, REPLACE, or APPEND.

Initiating the Batch Import

Before initiating the batch import, define all the input and output parameters you will use in the batch import by setting the values of macro variables or by assigning filerefs. After defining your parameters, initiate the batch import by executing the SASHELP.GISIMP.BATCH.SCL entry.

Note: You do not pass any parameters directly to the SCL entry. △

You can execute the SASHELP.GISIMP.BATCH.SCL entry with any one of the following methods:

- □ Enter the following command from a SAS command line:

```
AF C=SASHELP.GISIMP.BATCH.SCL
```

- □ Submit a DM statement that will issue the AF command:

```
DM 'AF C=SASHELP.GISIMP.BATCH.SCL';
```

- □ Use the CALL DISPLAY routine in an SCL program:

```
CALL DISPLAY('SASHELP.GISIMP.BATCH.SCL');
```

Examples of Batch Importing

Example 1: Batch Importing TIGER Files

This example imports the data for Wake County, North Carolina, from a 2006 Second Edition TIGER file, and appends the data for neighboring Durham County, North Carolina, from a separate TIGER file.

```
/* Define the input parameters for Wake County,
   North Carolina. */

   /* Define the import type. */
   %let IMP_TYPE = TIGER;

   /* Specify the complete path to the TIGER files for Wake County
     which were downloaded and unzipped using the required TIGER1
       and TIGER2 filerefs. */
   filename TIGER1 'tgr37183.rt1';
   filename TIGER2 'tgr37183.rt2';

/* Define the output parameters for Wake County,
   North Carolina. */
```

```
      %let MAPLIB = SASUSER;
      %let MAPCAT = TIGER;
      %let MAPNAME = COUNTIES;
      %let CATHOW = CREATE;
      %let SPALIB = SASUSER;
      %let SPANAME = COUNTIES;
      %let SPAHOW = CREATE;

   /* Initiate the batch import by executing the SCL entry. */
         DM 'AF C=SASHELP.GISIMP.BATCH.SCL';

   /* Define the input parameters for Durham County,
      North Carolina. */

       /* IMP_TYPE value stays the same, so you just
            need to reallocate the filerefs to point to the
           spatial data that was downloaded and unzipped
           for Durham County. */
       filename TIGER1 'tgr37063.rt1';
       filename TIGER2 'tgr37063.rt2';

   /* Define the output parameters for Durham County,
      North Carolina. */

       /* The locations will stay the same,
          so you only need to redefine CATHOW and
          SPAHOW to update the catalog entries and
          append the spatial data sets for the
          second import. */
       %let CATHOW = UPDATE;
       %let SPAHOW = APPEND;

   /* Initiate the batch import by executing the SCL
      entry a second time; this time to add the Durham
      County data to the Wake County data. */
      DM 'AF C=SASHELP.GISIMP.BATCH.SCL';
```

When the import completes, you can open the map named SASUSER.TIGER.COUNTIES. This map displays Wake and Durham counties.

Example 2: Batch Importing SASGRAPH and GENPOINT Data

This example creates a map of North Carolina with the state and county boundaries and then adds points at city locations. The state and county boundaries are imported from the MAPS library by using the SASGRAPH import type, and the points are appended using the GENPOINT import type.

```
   /* Construct the data sets to be imported into
      SAS/GIS.  The North Carolina state and county
      boundaries are obtained from the MAPS.USCOUNTY
      data set and the North Carolina city locations
      are obtained from the MAPS.USCITY data set.
      Both data sets are supplied with SAS/GRAPH
      software.  */
```

```
      /* Subset just the boundaries for the state of
         North Carolina. */
         data sasuser.nc;
            set maps.uscounty;
            /* 37 is the FIPS code for North Carolina.*/
            where state=37;
         run;

      /* Subset just the cities in North Carolina. */
         data sasuser.nccities;
            set maps.uscity;
            /* 37 is the FIPS code for North Carolina. */
            where state=37;
         run;

   /* Define the input parameters for the SASGRAPH
      import of the boundaries. */

      /* Define the import type. */
      %let IMP_TYPE = SASGRAPH;

      /* Specify where map data set is located. */
      %let INFILE=SASUSER.NC;

      /* Specify the identification variables, in
         hierarchical order (largest polygon first). */
      %let NIDVARS=2;
      %let IDVAR1=STATE;
      %let IDVAR2=COUNTY;

   /* Define the output parameters for the boundaries. */
      %let MAPLIB = SASUSER;
      %let MAPCAT = NC;
      %let MAPNAME = NC;
      %let CATHOW = CREATE;
      %let SPALIB = SASUSER;
      %let SPANAME = NC;
      %let SPAHOW = CREATE;

   /* Initiate the batch import by executing the SCL entry. */
      DM 'AF C=SASHELP.GISIMP.BATCH.SCL';

   /* Define the input parameters for the batch import of the
      GENPOINT data set for the cities.  */

      /* The import type has changed, so redefine the
         IMP_TYPE macro variable. */
      %let IMP_TYPE=GENPOINT;

      /* Specify where the generic point data is located. */
      %let INFILE=SASUSER.NCCITIES;

      /* Define the number of identification
         variables.  If you want all of the cities to
```

```
      be contained in one layer, don't define any. */
   %let NIDVARS=0;

/* Define the output parameters for the cities. */

   /* The locations will stay the same, so
      you only need to redefine CATHOW and SPAHOW to
      update the catalog entries and append the
      spatial data sets for the second import. */
   %let CATHOW = UPDATE;
   %let SPAHOW = APPEND;

/* Initiate the batch import by executing the SCL
   entry a second time; this time to add the points
   to the boundaries. */
   DM 'AF C=SASHELP.GISIMP.BATCH.SCL';
```

When the import completes, you can open the map named SASUSER.NC.NC. This map displays the state and county boundaries for North Carolina. You can choose to display the city points on the map.

File Reference Table for Batch Importing

Table 4.4 on page 51 lists the reserved filerefs for each of the different import types, a brief description of each file, and whether the fileref is required or optional when using that import type. For example, to import a Digital Line Graph file, you must allocate the file with a fileref of DLGIN.

Note: This information can also be found in the SASHELP.GISIMP data set. See "The SASHELP.GISIMP Data Set" on page 40 for more information. △

Table 4.4 File Reference Table for Batch Importing

IMP_TYPE Value	Fileref	File Description	Status
TIGER	TIGER1	TIGER basic data record (.rt1 file)	Required
	TIGER2	TIGER shape coordinate points (.rt2 file)	Required
	TIGER4	TIGER index to alternate feature names (.rt4 file)	Optional
	TIGER5	TIGER feature name list (.rt5 file)	Optional
	TIGER6	TIGER additional address and ZIP code (.rt6 file)	Optional
DYNAMAP	GDT1	Dynamap basic data record	Required
	GDT2	Dynamap shape coordinate points	Required
	GDT4	Dynamap index to alternate feature names	Optional
	GDT5	Dynamap feature name list	Optional
	GDT6	Dynamap additional address and ZIP code data	Optional
DLG	DLGIN	Digital Line Graph file	Required
DXF	DXFIN	DXF file	Required
ARC	ARCIN	Uncompressed ArcInfo interchange (E00) file	Required

IMP_TYPE Value	Fileref	File Description	Status
MAPINFO	MIF	MapInfo Interchange Format MIF file	Required
	MID	MapInfo Interchange Format MID file	Required

For the SASGRAPH and Generic import types, which require that the spatial data be stored in a SAS data set, do not allocate a reserved fileref to indicate the location of the data set. Instead, assign the name of the SAS data set to the INFILE macro variable.

Hints and Tips for Batch Importing

- The SAS/GIS Batch Import process provides less error checking than the SAS/GIS Spatial Data Importing window. Defining invalid values for the input parameters will cause the import to fail.
- The Importing window interface lets you modify the default composites and the default layer definitions before you proceed with the import. The Batch Import process does not provide this functionality. To modify the composites and layers before the import occurs, you must use the Importing window. However, whether you use the Importing window or the Batch Import process, you can always use PROC GIS after the import is complete to make changes to your map and its underlying components.
- The batch import for the TIGER and DYNAMAP import types does not automatically import all of the files, only the required file types 1 and 2. To use the Batch Import process to import any or all of the optional file types, you need to perform the following steps:

  1. Copy the SASHELP.GISIMP data set to a location where you have WRITE access. For example, copy the SASHELP.GISIMP data set into the SASUSER library:

     ```
     proc copy in=sashelp out=sasuser;
        select gisimp / mt=data;
     run;
     ```

  2. Edit the SASUSER.GISIMP data set and change the value of the REQ variable for the TIGER or DYNAMAP file type from 0 (zero) to 1. For example, if you have the TIGER4 and TIGER5 filerefs allocated to the TIGER file types 4 and 5, and you want them to be imported, then you could run the following DATA step to change the value of the REQ variable:

     ```
     data sasuser.gisimp;
        set sasuser.gisimp;
        /* Make sure that the values of the
           FILEREF variable are in all uppercase. */
        if fileref in ('TIGER4' 'TIGER5')
           then req=1;
     run;
     ```

     However, you could also use FSBROWSE, FSVIEW, VIEWTABLE, or any other method that you are familiar with to change the value of the REQ variable. Just remember that for the import type that you choose, it will import only from the filerefs for which REQ=1 in the data set.

3 For SAS/GIS to use the new SASUSER.GISIMP data set, you must define the USER_FIL macro variable to point to the name of the new data set. For example:

```
%let USER_FIL=SASUSER.GISIMP;
```

As long as the USER_FIL macro variable is defined when the SASHELP.GISIMP.BATCH.SCL entry is executed, it will use the current values of REQ to specify which files are imported.

- The SASHELP.GISIMP data set also contains the default librefs for the output catalog entries and spatial data sets. You can modify these defaults by making a copy of the SASHELP.GISIMP data set and changing the values for the DEFMLIB and DEFSLIB variables to a valid, assigned libref. You then need to define the macro variable USER_FIL to point to your modified copy of the data set.
- The data set that is specified by the USER_FIL macro variable is used by both the batch and interactive imports. If you have modified a copy of the SASHELP.GISIMP data set and assigned the data set name to the USER_FIL macro variable, but you do not want to have that data set used for a specific import, then redefine the USER_FIL macro variable to point to the default data set, SASHELP.GISIMP, before performing an import of either type.
- Throughout this chapter, all of the macro variable names, their values, and all filerefs have been shown in all uppercase for clarity. However, their names and values are not case-sensitive. For example, the following four statements are equivalent:

```
%let imp_type=sasgraph;
%let imp_type=SASGRAPH;
%let IMP_TYPE=sasgraph;
%let IMP_TYPE=SASGRAPH;
```

CHAPTER
5

Working with Spatial Data

SAS Data Sets

A SAS data set is a collection of data values and their associated descriptive information that is arranged and presented in a form that can be recognized and processed by SAS. SAS data sets can be data files or views. A SAS data file contains the following elements:

- data values that are organized into a rectangular structure of columns and rows
- descriptor information that identifies attributes of both the data set and the data values

A SAS view contains the following elements:

- instructions to build a table
- descriptor information that identifies attributes of both the data set and the data values

SAS data sets can be indexed by one or more variables, known as *key variables*. A SAS index contains the data values of the key variables that are paired with location identifiers for the observations that contain the variables. The value and identifier pairs are ordered in a B-tree structure that enables the engine to search by value. SAS indexes are classified as *simple* or *composite*, according to the number of key variables they contain.

For more information about SAS data sets, SAS files, SAS views, and SAS indexes, refer to *SAS Language Reference: Concepts*.

SAS/GIS Data Sets

As a component of SAS, SAS/GIS stores all of its data in SAS data sets. The SAS/GIS spatial database works as one logical entity, but is physically separated into six different categories of data sets:

- chains
- nodes
- details
- polygonal index
- label
- attribute

A given SAS/GIS map can reference only one chains, nodes, details, and label data set, but it can reference multiple polygonal index and attribute data sets. Multiple SAS/GIS maps can use a single set of chains, nodes, and details data sets.

Chains Data Set

The chains data set contains coordinates for the polylines that are used to form line and polygon features. A polyline consists of a series of connected line segments that are chains. A chain is a sequence of two or more points in the coordinate space. The end points, the first and last points of the chain, must be nodes. Each chain has a direction, from the first point toward the last point. The first point in the chain is the *from-node*, and the last point is the *to-node*. Relative to its direction, a chain has a left side and a right side. Points between the from-node and the to-node are *detail points*, which serve to trace the curvature of the feature that is represented by the chain. Detail points are not nodes.

The chains data set also lists the from-node and to-node row numbers in the nodes data set, as well as the number of detail points and the corresponding details data set row number. The left and right side attribute values (for example, ZIP codes and FIPS codes) are also stored in the chains data set.

Nodes Data Set

The nodes data set contains the coordinates of the end points for the chains in the chains data set and the linkage information that is necessary to attach chains to the correct nodes. A node is a point in the spatial data with connections to one or more chains. Nodes can be discrete points or the end points of chains. A node definition can span multiple records in the nodes data set, so only the starting record number for a node is a node feature ID.

Details Data Set

The details data set stores the curvature points of a chain between the two end nodes, which are also called the from-node and the to-node. That is, the details data set contains all the coordinates between the intersection points of the chain. The node coordinates are not duplicated in the details data set. The details data set also contains the chains data set row number of the associated chain.

Polygonal Index Data Sets

The polygonal index data set contains one observation for each polygon that was successfully closed during the index creation process. It is called a polygonal index because each observation is an index to a polygon in the chains data set. That is, it points to the starting chain in the chains data set for each of the polygons.

If polygon areas, perimeter distances, and centroid locations were computed, then that information is also stored in the polygonal index data set.

Label Data Set

The label data set defines the attributes of labels to be displayed on the map. The attributes include all of the information that is applicable for each label, such as location, color, size, source of the text for a text label, as well as other behavioral and graphical attributes.

Attribute Data Sets

Attribute data sets contain values related to the map features. The observations in attribute data sets must be associated with observations in the chains data set. Attribute data is used to display themes on the map and for spatially oriented reports, graphs, map actions, and so forth.

Managing Data Set Sizes

By their nature, spatial databases tend to be rather large. Users of spatial data want as much detail in the maps as they can get, which increases the demands on storage and processing capacity. Spatial data that is not carefully managed can become too large for easy use.

Here are five actions that you can take to manage the size of your spatial data sets. You need to perform most of these actions *before* importing your data into SAS/GIS.

- Reduce the spatial extent of the data.

  Do not store a larger area than you need. If you need a map containing one state, do not store a map containing all the states for a region. For example, if you need to work with a map of Oregon, do not store a map containing all of the Pacific Northwest.

- Store only the features that you need.

  If you do not need features such as rivers and lakes, do not store these features in your spatial data.

- Limit the amount of detail to what is necessary for your application.

  If you are using a map for which you do not require highly detailed boundaries, reduce the detail level and save storage space. If you are using SAS/GRAPH data sets, you can use the GREDUCE procedure in SAS/GRAPH software to reduce the detail level. If you are using a data set from another source, you'll have to reduce the level of detail before importing the data set into SAS/GIS.

- Reduce the number of attributes that are stored with the spatial data.

  If you do not need an attribute, and do not think you will ever need it again, delete it from your spatial data.

- Reduce the size of variables that are stored in the spatial data.

  Examine the method that you use for storing your variables and determine whether you can safely reduce the variable size that you use to store them.

  For example, if you have a numeric variable that contains a code that can be a maximum of two digits, perhaps it would be better to store it in a two-digit character variable rather than in an eight-byte numeric variable. Change the variables' defined types or lengths in a DATA step after you complete the import.

Of the five actions, reducing the number of attributes is the easiest to perform. Use the Import window, which you access by selecting **Modify Composites** from the GIS Spatial Data Importing window, to remove and drop unneeded composite variables from your data set as it is imported.

Import Type Specific Variables

The following tables describe the composites and variables that are created for each of the import types. All of the variables are located in the chains data set except for the X and Y variables, which are in the nodes data set.

Table 5.1 Partial Listing of Composites and Variables Specific to the ArcInfo Interchange Import Type

Composite	Variable 1	Variable 2	Type[1]	Description
ARCID	ARCIDL	ARCIDR	A or C	ARCID from the ArcInfo coverage. Maps made from line and point coverages do not have left and right variables.
ARCNUM			C	ARCNUM from the coverage.
'COVERAGE'	'COVERAGE'_L	'COVERAGE'_R	A or C	This variable is derived from the input filename. It is the last word preceding the file extension. For example, **/local/gisdata/montana.e00** would have a 'COVERAGE'[2] name of **montana**. The left variable would be **montanal**, the right variable would be **montanar**, and the composite type would be Area. Line and point coverages do not have left- and right-side variables, and the composite type would be Classification.
AREA	AREAL	AREAR	A	AREA from the coverage.
PERIMETER	PERIML	PERIMR	A	PERIMETER from the coverage.

Composite	**Variable 1**	**Variable 2**	**Type**[1]	**Description**
'ATTRIB'	'ATTRIB'L	'ATTRIB'R		All variables in the polygon, line, or point attribute tables are saved as composite variables. In the case of the polygon coverages, an L or an R is added to the end of the first five characters of the actual variable name.
COVER	_COVEL	_COVER	A or C	This variable contains the name stored in the *'COVERAGE'* variable.
SRC	_SRCL	_SRCR	C	Contains the string 'ARC'.
X	X		X	X coordinate.
Y	Y		Y	Y coordinate.

1 Values for Type are as follows:

A	Area
C	Classification
x	X coordinate
Y	Y coordinate

2 Names in single quotation marks, such as 'COVERAGE' and 'ATTRIB,' are GIS composite names.

Table 5.2 Partial Listing of Composites and Variables Specific to the Digital Line Graph (DLG) Import Type

Composite	**Variable 1**	**Variable 2**	**Type**[1]	Description
LMAJOR(**n**)	LMAJOR(**n**)		C	Major line attribute code.
LMINOR(**n**)	LMINOR(**n**)		C	Minor line attribute code.
NMAJOR(**n**)	NMAJOR(**n**)		C	Major node attribute code.
NMINOR(**n**)	NMINOR(**n**)		C	Minor node attribute code.
MAJOR(**n**)	AMAJORR(**n**)	AMAJORL(**n**)	A	Major area attribute code.
MINOR (**n**)	AMINORL(**n**)	AMINORR(**n**)	A	Minor area attribute code.
X	X		X	X coordinate.
Y	Y		Y	Y coordinate.

1 Values for Type are as follows:

A	Area
C	Classification
x	X coordinate
Y	Y coordinate

Table 5.3 Partial Listing of Composites and Variables Specific to the Drawing Interchange File (DXF) Import Type

Composite	Variable 1	Variable 2	Type[1]	Description
'ATTRIB'	'ATTRIB'L	'ATTRIB'R	A or C	All polygon, line, or point attributes are saved as composite variables. In the case of polygon maps, an **L** or **R** is added to the end of the first seven characters of the actual variable name.

1 Values for Type are as follows:

A	Area
C	Classification

Table 5.4 Partial Listing of Composites and Variables Specific to the Genline Import Type

Composite	Variable 1	Variable 2	Type[1]	Description
ID	ID		C	The ID variable from the data set.
'ATTRIB'	'ATTRIB'	'ATTRIB'	C	Any other variable in the data set is saved as a classification composite.
X	X		X	X coordinate.
Y	Y		Y	Y coordinate.

1 Values for Type are as follows:

C	Classification
x	X coordinate
Y	Y coordinate

Table 5.5 Partial Listing of Composites and Variables Specific to the Genpoint Import Type

Composite	Variable 1	Variable 2	Type[1]	Description
ID	ID		C	The ID variable from the data set.
'ATTRIB'	'ATTRIB'	'ATTRIB'	C	Any other variable in the data set is saved as a classification composite.
X	X		X	X coordinate.
Y	Y		Y	Y coordinate.

1 Values for Type are as follows:

C	Classification
x	X coordinate
Y	Y coordinate

Table 5.6 Partial Listing of Composites and Variables Specific to the MapInfo Import Type

Composite	Variable 1	Variable 2	Type[1]	Description
'ATTRIB'	'ATTRIB'L	'ATTRIB'R	A or C	All polygon, line, or point attributes are saved as composite variables. In the case of polygon maps, an **L** or **R** is added to the end of the first seven characters of the actual variable name.
LINELYR			C	This variable is derived from the input filename. It is the last word preceding the file extension. For example, **/local/gisdata/montana.mif** would have a LINELYR name of **montana**.
PTLYR			C	This variable is derived from the input filename. It is the last word preceding the file extension. For example, **/local/gisdata/montana.mif** would have a PTLYR name of **montana**.
POLYLYR			A	This variable is derived from the input filename. It is the last word preceding the file extension. For example, **/local/gisdata/montana.mif** would have a POLYLYR name of **montana**.
'MAP'	'MAP'L	'MAP'R	A or C	This variable is derived from the input filename. It is the last word preceding the file extension. For example, **/local/gisdata/usa.mif**, would have a 'MAP' name of **usa**. The left variable would be **usal**, the right variable would be **usar** and, in this case, the composite type would be Area. Line and point maps do not have left- and right-side variables, and the composite would be Classification.

1 Values for Type are as follows:

A	Area
C	Classification

Table 5.7 Partial Listing of Composites and Variables Specific to the SAS/GRAPH and Genpoly Import Types

Composite	Variable 1	Variable 2	Type[1]	Description
'IDVAR'n	'IDVAR'L	'IDVAR'R	A	An area composite variable is created for each ID variable (IDVAR) selected by the user in the **ID vars** list box. In the case of polygon maps, an **L** or **R** is added to the end of the first seven characters of the actual variable name.

1 Values for Type are as follows:

A	Area

Table 5.8 Composites and Variables Specific to the TIGER and DYNAMAP Import Types

Composite	Variable 1	Variable 2	Variable 3	Variable 4	Type[1]	Description
ADDR	FRADDL	FRADDR	TOADDL	TOADDR	ADDR	Address range.
BLOCK	BLOCKL	BLOCKR			A	Block number.
CFCC	CFCC				C	Feature classification code.
COUNTY	COUNTYL	COUNTYR			A	County FIPS code.
DIRPRE	DIRPRE				ADDRP	Feature direction prefix.
DIRSUF	DIRSUF				ADDRS	Feature direction suffix.
FEANAME	FEANAME				C	Feature name.
MCD	MCDL	MCDR			A	Minor civil division.
PLACE	PLACEL	PLACER			A	Incorporated place code.
RECTYPE	RECTYPE				C	Record type.
STATE	STATEL	STATER			A	State FIPS code.
TRACT	TRACTL	TRACTR			A	Census tract.
ZIP	ZIPL	ZIPR			A	ZIP code.
BG	BGL	BGR			A	Block group.
LONGITUDE	X				X	Longitude.
LATITUDE	Y				Y	Latitude.

1 Values for Type are as follows:

A	Area
ADDR	Address
ADDRP	Address Prefix
ADDRS	Address Suffix
C	Classification
x	Longitude
Y	Latitude

Data Set and Catalog Entry Interactions

SAS/GIS software uses SAS catalog entries to store *metadata* for the spatial database, that is, information about the spatial data values in the spatial data sets. SAS/GIS spatial databases use the following entry types:

- spatial entries
- coverage entries
- layer entries
- map entries
- composites

Spatial Entries

A *spatial entry* is a SAS catalog entry of type GISSPA that identifies the spatial data sets for a given spatial database and defines relationships between the variables in those data sets.

Spatial entries are created and modified using the SPATIAL statement in the GIS procedure.

Note: You can also create a new spatial entry by selecting the following from the GIS Map window's menu bar:
File ▶ Save As ▶ Spatial △

SAS/GIS software supports *simple spatial entries* and *merged spatial entries*.

Simple Spatial Entries

Simple spatial entries contain the following elements:

- references to the chains, nodes, and details data sets that contain spatial information.
- references to any polygonal index data sets that define the boundaries of area features in the spatial data.
- definitions for composites that specify how the variables in the spatial data sets are used. See "Composites" on page 67 for more information about composites.
- the definition for a lattice hierarchy that specifies which area features in the spatial data enclose or are enclosed by other features.
- parameters for the projection system that is used to interpret the spatial information that is stored in the spatial data sets.
- the accumulated bounding extents of the spatial data coordinates of its underlying child spatial data sets, consisting of the minimum and maximum X and Y coordinate values and the ranges of X and Y values.

Merged Spatial Entries

Merged spatial entries have the following attributes:

- consist of multiple SAS/GIS spatial databases that are linked together hierarchically in a tree structure.
- contain logical references to two or more child spatial entries. A child spatial entry is a dependent spatial entry beneath the merged spatial entry in the hierarchy.
- contain specifications of how the entries were merged (by overlapping or edge matching).

- do not have their own spatial data sets.
- reference the spatial data sets that belong to the child spatial entries beneath them on the hierarchy.
- do not have references to any polygonal index data sets that define the boundaries of area features in the spatial data.
- do not have definitions for composites that specify how the variables in the spatial data sets are used. See "Composites" on page 67 for more information about composites.
- do not have the definition for a lattice hierarchy that specifies which area features in the spatial data enclose or are enclosed by other features.
- do not have parameters for the projection system that is used to interpret the spatial information stored in the spatial data sets.
- contain the accumulated bounding extents of the spatial data coordinates of their underlying child spatial entries, consisting of the minimum and maximum X and Y coordinate values and the ranges of X and Y values.

Merged spatial entries can help you to manage your spatial data requirements. For example, you might have two spatial databases that contain the county boundaries of adjoining states. You can build a merged spatial entry that references both states and view a single map that contains both states' counties. Otherwise, you would have to import a new map containing the two states' counties. This new map would double your spatial data storage requirements.

The following additional statements in the GIS procedure update the information in the spatial entry:

COMPOSITE statement
: creates or modifies composites that define the relation and function of variables in the spatial data sets. The composite definition is stored in the spatial entry. See "COMPOSITE Statement" on page 94 for details about creating or modifying composites.

POLYGONAL INDEX statement
: updates the list of available index names stored in the spatial entry. See "POLYGONAL INDEX Statement" on page 99 for details about creating or modifying polygonal indexes.

LATTICE statement
: updates the lattice hierarchy stored in the spatial entry. See "LATTICE Statement" on page 102 for details about defining lattice hierarchies.

You can view a formatted report of the contents of a spatial entry by submitting a SPATIAL CONTENTS statement in the GIS procedure.

See "SPATIAL Statement" on page 90 for details about using the GIS procedure to create, modify, or view the contents of spatial entries.

Coverage Entries

A *coverage entry* is a SAS catalog entry of type GISCOVER that defines the subset, or *coverage*, of the spatial data that is available to a map. SAS/GIS maps refer to coverages rather than directly to the spatial data.

A coverage entry contains the following elements:

- a reference to the root spatial entry.
- a WHERE expression that describes the logical subset of the spatial data that is available for display in a map.

Note: The expression WHERE='1' can be used to define a coverage that includes all the data that is in the spatial database. WHERE='1' is called a *universal coverage*. △

The WHERE expression binds the coverage entry to the spatial data sets that it subsets. The WHERE expression is checked for compatibility with the spatial data when the coverage entry is created and also whenever a map that uses the coverage entry is opened.

- the maximum and minimum X and Y coordinates in the portion of the spatial data that meets the WHERE expression criteria for the coverage.

  These maximum and minimum coordinates are evaluated when the coverage is created. The GIS procedure's COVERAGE CREATE statement reads the matching chains and determines the extents from the chains' XMIN, YMIN, XMAX, and YMAX variables. If you make changes to the chains, nodes, and details data sets that affect the coverage extents, use the COVERAGE UPDATE statement to update the bounding extent values.

Multiple coverage entries can refer to the same spatial entry to create different subsets of the spatial data for different maps. For example, you could define a series of coverages to subset a county into multiple sales regions according to the block groups that are contained in each of the regions. The spatial data for the entire county would still be in a single spatial database that is represented by the chains, nodes, and details data sets and by the controlling spatial entry.

Coverage entries are created and modified by using the COVERAGE statement in the GIS procedure. You can view a formatted report of the contents of a coverage entry by submitting a COVERAGE CONTENTS statement in the GIS procedure. (The contents report for a coverage entry also includes all the contents information for the root spatial entry as well.)

See "COVERAGE Statement" on page 105 for more information about creating, modifying, or viewing the contents of coverage entries.

Layer Entries

A *layer entry* is a SAS catalog entry of type GISLAYER that defines the set of features that compose a layer in the map. A layer entry contains the following elements:

- a WHERE expression that describes the common characteristic of features in the layer.

  The WHERE expression binds the layer entry to the spatial data even though the WHERE expression is stored in the layer entry. The layer is not bound to a specific spatial entry, just to those entries that represent the same type of data. Therefore, a layer that is created for use with data that is imported from a TIGER file can be used with data that is imported from any TIGER file; however, not all file types can take advantage of this behavior. The WHERE expression is checked for compatibility with spatial data when the layer entry is created and also whenever a map that uses the layer entry is opened.

  Note: When you define area layers, you can specify a composite as an alternative to specifying an explicit WHERE expression. However, the layer entry stores the WHERE expression that is implied by the composite. For example, if you specify STATE as the defining composite for a layer, and the STATE composite specifies the following variables: VAR=(LEFT=STATEL,RIGHT=STATER), then the implied WHERE expression that is stored in the layer entry is 'STATEL NE STATER'. △

- option settings for the layer such as the layer type (point, line, or area), whether the layer is static or thematic, whether it is initially displayed or hidden, whether detail points are drawn for the layer, and the scales at which the layer is automatically turned on or off.

- the graphical attributes that are necessary to draw the layer when it is displayed as a static layer.
- the attribute links, theme range breaks, and graphical attributes if the layer contains themes.

See "LAYER Statement" on page 107 for more information about creating, modifying, or viewing the contents of layer entries.

Map Entries

A *map entry* is a SAS catalog entry of type GISMAP. Map entries are the controlling entries for SAS/GIS maps because they tie together all the information that is needed to display a map. A map entry contains the following elements:

- a reference to the coverage entry that defines the subset of the spatial data that is available to the map. Note that the map entry refers to a particular coverage of the spatial data rather than directly to the spatial entry.
- references to the layer entries for all layers that are included in the map.
- references to any attribute data sets that are associated with the map, for example, the data sets that are used for the map actions, along with definitions of how the attribute data sets are linked to the spatial data.
- a reference to the SAS data set that contains labels for map features.
- definitions for the actions that can be performed.
- definitions for map legends.
- parameters for the projection system that is used to project the spatial data coordinates for display.
- option settings for the map, including the following:
  - the units and mode for the map scale
  - whether coordinate, distance, and attribute feedback are displayed
  - whether detail points are read
  - whether the tool palette is active.

Map entries are created by using the MAP CREATE statement in the GIS procedure. However, much of the information that is stored in the map entry is specified interactively in the GIS Map window.

You can view a formatted report of the contents of a map entry by submitting a MAP CONTENTS statement in the GIS procedure. (The contents report for a map entry includes all the contents information for the spatial, coverage, and layer entries as well.)

See "MAP Statement" on page 129 for details about using the MAP statement. See Chapter 10, "SAS/GIS Windows" in the SAS/GIS Software: Usage and Reference, Version 6, for details about the items that can be specified interactively in the GIS Map window.

Composites

For most operations that involve the spatial database, you refer to composites of the spatial data variables rather than directly to the variables in the spatial data sets. A composite consists of the following elements:

- a *variable association* that identifies which variable or variables in the spatial database comprise the association. The variable association can specify a single variable, a pair of variables that define a bilateral (left-right) association, or two pairs of variables that define the start and end of a directional (from-to) bilateral association.
- a *class attribute* that identifies the role of the composite in the spatial database.

For example, if the chains data set has a variable named FEANAME that contains feature names, you can create a composite for the FEANAME variable that assigns the class attribute NAME to indicate that the association represents feature names. Or, if the chains data set has COUNTYL and COUNTYR variables that contain the codes for the counties on the left and right sides of the chains, you can create a composite named COUNTY. The composite identifies the bilateral relationship between these two variables and assigns the class attribute AREA to indicate that the association defines county areas in the spatial data.

Composites are created and modified using the COMPOSITE statement in the GIS procedure. Composite definitions are stored in the spatial entry. When a spatial action is performed in a map, the variables referenced by composites for the selected map features are displayed in the Spatial Information window.

See “COMPOSITE Statement” on page 94 for more information about creating or modifying composites.

Merging Spatial Data with the MERGE= Argument

Overview

MERGE= is an argument of the GIS procedure's SPATIAL statement that enables you to build a new spatial entry by referencing two or more existing spatial entries. The dependent data sets for the spatial entries are not actually combined when you use the MERGE= argument; the new spatial entry includes them by reference.

Syntax

MERGE=(<*libref.catalog.*>*spatial-entry-1* <, ..., <*libref.catalog.*>*spatial-entry-n*>)
<EDGEMATCH | OVERLAP>

Note: Keep in mind that MERGE is specified as an option on a SPATIAL statement. △

If you specify a one-level name for any of the entries to be merged, the spatial entry is assumed to be in the catalog that is specified in the CATALOG= argument with the PROC GIS statement or in the most recently issued CATALOG statement. An error occurs if you have not specified a catalog before specifying the names of the entries you want to merge.

Types of Merge Operations

The MERGE= argument accepts the following arguments:

EDGEMATCH
: locates common boundaries between the merged spatial entries and updates missing left- or right-side composite variable values in the chains data that lie on the boundaries.

In other words, the EDGEMATCH operation compares the chains in the different data sets and finds those chains that map the same feature. When it finds the same chain in both data sets, it replaces any missing left- or right-side composite values in either chain with the valid values from the other data set. EDGEMATCH also creates a merged spatial entry that references other spatial entries (either merged or simple) that you specified with the MERGE= argument.

EDGEMATCH rewrites the specified chains data sets. You cannot reverse this operation.

OVERLAP
: merges spatial entries without attempting to match boundaries. OVERLAP is the default behavior of the MERGE= argument. The OVERLAP argument creates a merged spatial entry that references the specified spatial entries (either merged or simple).

OVERLAP does not rewrite the specified chains data sets.

For more information, see "SPATIAL Statement" on page 90.

Benefits of Merging Data

Merging data enables you to construct maps that show larger geographic areas without the overhead of storing duplicate spatial data sets. For example, you might have chains, nodes, and details data sets for each U. S. state. If you want to create a map of New England, you do not have to physically combine and duplicate the individual data sets for the six states composing the region. Instead, you can create a merged spatial entry named `New_England` that references the individual states' simple spatial entries.

Edge matching provides a mechanism to update adjoining spatial data sets to replace missing left or right values in the chains data sets. Using the New England example, the chains in the chains data set for New Hampshire that lie along the Vermont border contain the FIPS code of 33 on one side of each chain. The other side of each chain has a missing value. The corresponding chains in the Vermont chains data set contain the Vermont FIPS code of 50 on one side and a missing value on the other side. An edge match merge of the two data sets locates these common boundary chains in each data set and replaces the missing values with the correct FIPS code for the adjoining state. It will also create a merged spatial entry that references the New Hampshire and Vermont simple spatial entries.

The EDGEMATCH operation creates a single merged spatial entry by which you can create a map of the two states. It also adds the Vermont FIPS code to the appropriate chains in the New Hampshire data set, and adds the New Hampshire FIPS code to the corresponding chains in the Vermont data set.

Sample SAS/GIS Spatial Database

SAS/GIS offers a code sample that creates a fully functional SAS/GIS spatial database. This sample is available in the online Help. In the GIS Map window, select **Help ▶ Getting Started with SAS/GIS Software ▶ Create Data**. You can use this sample map with the SAS/GIS interface and the GIS procedure.

Hints and Tips for Working with Spatial Data

- When SAS/GIS uses a coverage that is not universal—that is, one in which the value of the WHERE expression is not '1'—to subset a map, all of the layers in the map must also satisfy this WHERE expression. If any of the layers do not satisfy this WHERE expression, some features of the map might not be displayed, and the reason might not be apparent.

  For example, if you have a map of the United States and you want to create a subset map containing just North Carolina and Virginia, you could use the following COVERAGE statement to create the subset map:

  ```
  COVERAGE CREATE NCVA /
      where='STATEL IN(37 51) OR
             STATER IN(37 51)';
  ```

  Any points or lines that do not have 37 or 51 as the STATEL or STATER value will not display on the map.

- Defining a layer with WHERE='1' displays all of the features in the underlying spatial data that have that type. For example, if you have a map with a point layer that contains capital cities, and you add a new point layer for grocery store locations by using WHERE='1' for the layer definition, the grocery store layer will display all of the point features in the spatial data. This layer includes capital cities, grocery stores, and all other point features in the spatial data. You might find this confusing if you are not aware that all point features are being displayed when you intend to display only one layer.

  You might encounter this situation because the GENPOINT import, by default, defines all point layers with a WHERE='1' expression. You can click the `Modify layers` button on the GIS Spatial Data Importing window to redefine the layer definition to be a WHERE expression that uniquely identifies the set of points in the layer. If the layer already exists on the map, you can use the LAYER statement in the GIS procedure to redefine the layer with a WHERE expression that defines only those points in the layer.

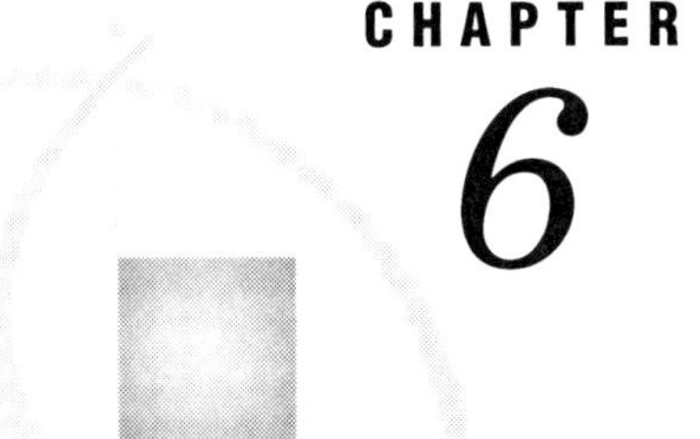

CHAPTER 6

Batch Geocoding

Overview of Batch Geocoding

Geocoding is the process of adding location information to an existing data set that contains address data. *Location information* is the X and Y coordinate data for the street addresses on a map. The geocoding facility in SAS/GIS software attempts to match each address in a SAS address data set to a location on a map. If a match is found, the X and Y coordinates of the address are added to the address data set. Other attribute information about the matched location can also be added to the address data set.

SAS/GIS software provides an interactive interface for geocoding through the GIS Geocoding window. The window is convenient for geocoding individual address data sets. However, if you have a large number of data sets or a data set with a large number of observations that you want to geocode, you might find the *batch geocoding facility* to be more convenient. The batch geocoding facility in SAS/GIS allows data to be geocoded without invoking SAS/GIS, without user intervention, and with improved performance. For example, you can set up a program to run overnight to geocode address data sets without user interaction.

Addresses in Spatial Data

In order to use geocoding in SAS/GIS, your spatial database must contain address information. SAS/GIS uses CLASS values for composites to identify address information in the spatial database. One of the composites must be defined as `CLASS=CITY` to indicate the city name, and one of the composites must be defined as `CLASS=ADDRESS` to indicate the location portion of the address. Composites that are defined with other CLASS values, such as ZIP, serve to improve accuracy. You use the COMPOSITE CREATE statement in the GIS procedure to add address-related composites to the spatial entry. You can use the SPATIAL CONTENTS statement in the GIS procedure to view the composites that are defined for your spatial database.

The following composite CLASS values identify elements of the address information:

NAME
: identifies the name component of the address feature, such as `Main` in the address `101 N Main Ave`.

TYPE
: identifies the type component of the address feature, such as `Ave` in the address `101 N Main Ave`.

ADDRESS
: identifies the specific address of the feature, such as `101` in the address `101 N Main Ave`. This composite is required when doing geocoding.

  A chain has four values to define the address range for each side:

FROMLEFT	beginning address on the left side.
TOLEFT	ending address on the left side.
FROMRIGHT	beginning address on the right side.
TORIGHT	ending address on the right side.

DIRECTION_PREFIX
: identifies the directional prefix component of the address feature, such as `N` in the address `101 N Main Ave`.

DIRECTION_SUFFIX
: identifies the directional suffix component of the address feature, such as `W` in the address `1141 First St W`.

CITY | PLACE
: identifies the value as a city name. This composite is required when doing geocoding.

STATE
: identifies the value as a state name.

ZIP
: identifies the value as a ZIP code value.

PLUS4
: identifies the value as a ZIP+4 extended postal code value.

You can use the SPATIAL CONTENTS statement in the GIS procedure to determine whether your spatial database contains the minimum composites that are necessary to perform geocoding. Submit the following statements in the SAS Program Editor for the spatial entry that you want to geocode against:

```
proc gis catalog=libref.catalog;
spatial contents spatial-entry;
run;
quit;
```

The output that is produced by the SPATIAL CONTENTS statement will include a list of all of the composites that are defined for the specified spatial entry. If the spatial database includes address information, this list will include some or all of the composites that are defined with the required CLASS values for address information.

Using Batch Geocoding

Using the batch geocoding facility is a two-step process:

1 Use the %GCBATCH macro to assign values to the macro variables that control the geocoding process. See "%GCBATCH Macro Statement" on page 77 for more information.

2 Call the SCL program to perform batch geocoding, SASHELP.GIS.GEOCODEB.SCL. In a SAS program, you can use the DM statement to issue an AF command to execute the SCL, as follows:

```
dm 'af c=sashelp.gis.geocodeb.scl; run;';
```

Note: If you are invoking SCL from your own frame application, you must use CALL DISPLAY instead of the DM command, for example, **call display('sashelp.gis.geocodeb.scl')**. △

How Batch Geocoding Works

To achieve the most accurate geocoding, ensure that the address data set to be geocoded contains name, address, city, state, ZIP code, and ZIP+4 variables. At least the address and city variables are required.

Created Data Sets

The geocoding facility first reads the chains, nodes, and details data sets for the map specified in the %GCBATCH macro. Then it creates new data sets for the sorted and summarized versions in the SAS library that was specified with the GLIB macro variable. Names for the geocoding data sets are generated from the specified map's chains data set name. For example, if your chains data set is GMAPS.USAC and you specify **GLIB=GEOLIB** in the %GCBATCH macro, then the geocoding facility creates the following data sets:

GEOLIB.USAS
: contains sorted chains.

GEOLIB.USAM
: contains matchable street data summarized from the chains data set and sorted by state, ZIP code, street name, and city.

GEOLIB.USAP
: contains point coordinates along the street segment taken from the map's nodes and details data sets.

These summary data sets are created automatically before the first address-matching process begins. After the data sets are created, they are regenerated only when the map's chains data set is updated or when NEWDATA=YES is specified in the %GCBATCH macro.

When choosing the SAS library to use for these created data sets, consider that—depending on the area of the base map—they can be quite large. If you use the WORK library, then the data sets will be deleted at the end of the current SAS session and must be regenerated if you want to perform geocoding again in a future SAS session.

Reference Data Sets

Additional data sets used in geocoding are supplied by SAS:

SASHELP.GCTYPE
: contains the official street abbreviations used by the U.S. Postal Service and in TIGER data from the U.S. Census Bureau. These values are used to standardize your address observations before geocoding.

SASHELP.PLFISP
: contains place names, state codes, and FIPS place codes for U.S. locations. The places primarily represent cities and towns, but the data set also includes some national parks, industrial parks, military installations, and so forth.

SASHELP.ZIPCODE
: contains U.S. ZIP codes, FIPS state and city codes, city names, and post office names. The ata set also contains the latitude and longitude for the centroid of each ZIP code area. If an address is not matched in the primary geocoding data sets, this data set is searched for a matching ZIP code. Updates for this data are available from the SAS Maps Online area at http://support.sas.com.

Match Addresses

The geocoding facility uses these data sets to match the addresses in the address data set. As it is processing the address data set, the geocoding facility provides a progress indicator. For every 10 percent of the addresses that are geocoded, a message is written to the SAS log.

When a match is found, the coordinates of the address location are added to the address data set, along with any other composite values for the specified address. For example, if the spatial data has a composite named TRACT that contains census tract numbers, you can use the geocoding process to add a TRACT variable to your address data set. The resulting geocoded address data set can be used as attribute data for the map, or it can be imported to add point data to the map by using a generic import.

If an address cannot be matched to the spatial data but the address includes a ZIP code, then the X and Y coordinates of the center of the ZIP code centroid for the zone are returned instead of the exact coordinates of the address. The centroid coordinates are read from the SASHELP.ZIPCODE data set.

For matching purposes, the geocoding process converts the address components to uppercase and attempts to convert direction and street type values to standard forms. The standardized versions of the address components are also added to the address data set. The M_ADDR, M_CITY, M_STATE, M_ZIP, and M_ZIP4 variables that are added to the address data set reflect the address values that were actually matched during the geocoding process. If a matching observation was found in the sorted chains data set, that row number is placed in the M_OBS variable.

Address Match Scoring

All address matches are not equal. The geocoding process attempts to match different elements of each specified address. When multiple address elements match, the resulting X/Y location is more certain. The geocoding process adds _SCORE_, _STATUS_, and _NOTES_ variables to the address data set to indicate which elements were matched. These variable values can also indicate whether there was a problem with a specific part of the address.

The _SCORE_ variable's value is a numeric rating of the certainty of the address match. A higher score indicates a better match. The score is calculated by adding points for matching various components of the address.

A score of 100 indicates that a match was found for all of the components of the address. A score of 100 is possible only if the address in the data set includes values for all components and the geocoding lookup data contains variables for all components. For example, if the address in the data set does not have a ZIP+4 value or if the lookup data set does not have a PLUS4 type variable, then the highest possible score is 95.

Table 6.1 _SCORE_ Values for Address Elements

Address Element Matched	Value added to _SCORE_ Value
Street number	40
Street name	20
Street type	5
Street direction	5
City	5
State	5
Five-digit ZIP code	15
First three digits of ZIP code	5
ZIP+4 code	5

The _STATUS_ variable provides a general indication of the match result:

Table 6.2 _STATUS_ Values for Match Results

STATUS Value	Description
found	Street name and ZIP code or city and state match found. X/Y interpolated along street. _SCORE_ indicates how many elements were matched.
ZIP Match	Street name not found in lookup data. ZIP code was found in SASHELP.ZIPCODE. X/Y for ZIP center is within the lookup data extents.
ZIP Match OffMap	Street name not found in lookup data. ZIP code was found in SASHELP.ZIPCODE. X/Y for ZIP center is outside lookup data extents.
City/State Match	Street name not found in lookup data. City and state elements found in SASHELP.ZIPCODE. Multiple city and state matches were averaged for X/Y.
City not found	Address had missing ZIP code value. City is not in SASHELP.ZIPCODE. X/Y values are missing.
Unknown Address	No part of address was matched. X/Y values are missing.

The _NOTES_ variable provides additional details on which address elements were matched or invalid:

Table 6.3 _NOTES_ Values for Match Results

NOTES Value	Description
ZC	Five-digit ZIP code matched.
ZC3	First three digits of ZIP code matched.
AD	Street name matched.
TY	Street type matched.
DP	Street direction prefix matched.
DS	Street direction suffix matched.
NM	House number matched.
ST	State matched.
CT	City matched.

NOTES Value	Description
CT3	Used with ZC3. Street matched only first 3-digits of the ZIP code in lookup data and either the city value was missing in the address or the city and state pair in lookup data differed.
NOADD	Street address is invalid.
NOZC	Address ZIP code is missing.
NOCT	Address city name is invalid.

%GCBATCH Macro Statement

The %GCBATCH macro sets the input parameters for the batch geocoding program. The macro accepts the following information:

- □ the name of the address data set to geocode
- □ the variable names in the address data set
- □ the name of the map entry
- □ the library in which to store geocoding data sets, and whether new copies of the data sets are created
- □ the name of an alternate ZIP code centroids data set that is used instead of SASHELP.ZIPCODE
- □ the names of any additional polygonal composites to add to the address data set.

The %GCBATCH macro has the following general form. The elements within the %GCBATCH macro can appear in any order, but they must be separated by commas.

%GCBATCH(

<GLIB=*geocoding-library*,>

<ZIPD=*ZIP-centroids-data-set*,>

GEOD=*address-data-set*,

<NV=*name-var*,>

AV=*address-var*,

CV=*city-var*,

<SV=*state-var*,>

<ZV=*ZIP-var*,>

<P4V=*ZIP+4-var*,>

MNAME=*map-entry*,

<PV=*area-composite-list*,>

<NEWDATA=YES | NO>);

where

AV=*address-var*
: specifies the name of the variable that stores the complete street address in the address data set that you want to geocode. This includes the house number, street name, and street type (for example, 3922 Oak Avenue). This parameter is required.

CV=*city-var*
: specifies the name of the variable that stores the city name portion of the address in the address data set that you want to geocode. This parameter is required.

GEOD=*address-data-set*
: specifies the address data set that you want to geocode. This parameter is required.

GLIB=*geocoding-library*
: specifies the libref for the SAS library where all of the sorted and summarized chains, nodes, and details data sets that are created for the geocoding process are stored. This parameter is optional.

Note: The SAS library that you specify for the *GLIB=* argument should be on a volume that has a large amount of free space because the geocoding data sets might be quite large. Also, to take full advantage of the geocoding facility, you should specify a permanent SAS library. The default for this variable is WORK, but data sets in the WORK library are deleted when the SAS session is terminated, so the geocoding data sets are lost. If geocoding data sets already exist in the specified library at the start of the geocoding process, the geocoding facility checks their creation dates against the creation date of the chains data set. The geocoding data sets are created again only if the chains data set has a more recent creation date. The first time that you geocode with a particular chains data set, the process takes considerably longer because these geocoding data sets are being created, sorted, and indexed. Subsequent geocoding times, however, are much faster as long as the parent chains data set has not been modified. △

MNAME=*map-entry*
: specifies the name of the GISMAP entry for the SAS/GIS spatial database (the chains, nodes, and details data sets) that you are using for geocoding. The geocoding process uses the projection information in the map entry to ensure that the X and Y coordinates that are returned for the address will be in the same coordinate system as the spatial data for the map. The MNAME= argument should use the form *libref.catalog-name.entry-name*. This parameter is required.

NEWDATA=YES | NO
: specifies whether the geocoding lookup data sets are created again if they already exist. The default is NEWDATA=NO. If you set NEWDATA=NO, the geocoding facility searches the SAS library that you specified with the GLIB macro variable for geocode data sets that were created for the spatial entry. The geocoding facility checks the creation date of existing geocode data sets against the creation date of the spatial entry. If the creation date of the geocode data sets is more recent than the creation dates of the spatial entry, the geocoding facility uses the geocode data sets. Otherwise it creates new geocode data sets.

Use NEWDATA=YES to force the geocoding facility to build new versions of the geocoding data sets. You should specify NEWDATA=YES if the existing geocoding data sets were created with an earlier version of SAS/GIS software. This parameter is optional.

NV=*name-var*
: specifies the name of the variable that stores the name portion of the address in the address data set that you want to geocode.

This parameter is optional.

PV=*area-composite-list*
: specifies the list of polygonal (area) composite values that you want added as variables to the address data set along with the X and Y coordinates of the address. By default, no other variables are added. Use spaces to separate composite names in the list. For example, the following specification adds the county and census tract and block values along with the address coordinates:

```
pv=county tract block,
```

This parameter is optional.

P4V=*ZIP+4-var*
: specifies the name of the variable that stores ZIP+4 postal codes in the address data set that you want to geocode.

This parameter is not required, but the accuracy of the geocoding process might be reduced if you omit it.

SV=*state-var*
: specifies the name of the variable that stores the state or province name portion of the address in the address data set you want to geocode.

This parameter is not required, but the accuracy of the geocoding process might be reduced if you omit it.

ZIPD=*ZIP-centroids-data-set*
: specifies a data set that contains the coordinates of the centers of ZIP code zones. (If an address includes a ZIP code and the street address cannot be matched, the geocoding facility supplies the ZIP code centroid coordinate instead of the address coordinate.) The default is ZIPD=SASHELP.ZIPCODE, which specifies the SASHELP.ZIPCODE data set that is supplied with SAS software. Updated data sets are available from the SAS Maps Online area at http://support.sas.com. This parameter is optional.

ZV=*ZIP-var*
: specifies the name of the variable that stores the ZIP code portion of the address in the address data set that you want to geocode.

This parameter is not required, but the accuracy of the geocoding process might be reduced if you omit it.

Batch Geocoding Example

The following example uses the batch geocoding macro to geocode an address data set using a copy of the MAPS.WAKE.TRACT map supplied with SAS/GIS software. That map was originally created by importing the U.S. Census Bureau TIGER files for Wake County, North Carolina. This example uses a copy in the WORK library rather than the original in the MAPS library to show how the geocoded addresses can be imported and appended to the spatial data.

```
/*--- Copy the base map to the WORK library ---*/
proc gis;
   copy MAPS.WAKE.TRACT.GISMAP / /* Map entry to copy */
           destlib = WORK        /* Destination library */
           destcat = WORK.WAKE   /* Destination catalog */
           sel     = (_all_)     /* Copy all map components */
           blank                 /* Clear internal map path */
           replace;              /* Overwrite existing entry */
quit;
```

```
/*--- Create the address data set to geocode ---*/
data WORK.ADDRESSES (label='Data set to geocode');
   input address  $ 1-23    /* Street address */
         resident $ 24-48   /* Person at the location */
         zip      $ 49-53   /* 5-digit US postal code */
         city     $ 55-69   /* City name */
         state    $ 70-71;  /* US state name */
cards;
700  Madison Avenue    Patricia Smith           27513 Cary
506  Reedy Creek Road  Jean Francois Dumas      27513 Cary
1106 Medlin Drive      Michael Garriss          27511                NC
1150 Maynard Road      Kaspar Gutman            27511 Cary
138  Dry Ave.          Susan Lang               27511                NC
3112 Banks Road        Roy Hobbs                27603 Raleigh        NC
305  Mill Creek Drive  Alan Picard              27526 Fuquay-Varina  NC
1998 S. Main St.       Guillermo Ugarte               Wake Forest
7825 Old Middlesex Rd  Capt. Jeffrey Spaulding  27807 Bailey         NC
5550 Old Stage Road    Emily Joyner             27603 Raleigh        NC
3212 Avent Ferry Road  Fred C. Dobbs            27540                NC
1050 King Charles Rd.  Karin Schmidt                . Raleigh        NC
6819 Buffaloe Road     Ferdinand Paulin         27604                NC
3211 Constant Circle   Gordon Miller            34121
6111 Old Faison Road   Alan Picard              27545 Knightdale
725  N. Raleigh Street Evan Rudde               27501 Angier         NC
;
run;

/*--- Set up variables for the Batch Geocoding program ---*/
%gcbatch( glib  = WORK,              /* Geocoding library */
          geod  = WORK.ADDRESSES,    /* Address data to geocode */
          nv    = RESIDENT,          /* Who's at the address */
          av    = ADDRESS,           /* Address variable */
          cv    = CITY,              /* Place name */
          sv    = STATE,             /* State name */
          zv    = ZIP,               /* ZIP code (5-digit) */
          pv    = TRACT,             /* AREA value from map data */
          mname = WORK.WAKE.TRACT); /* Map data used for geocoding */

/*--- Run the Batch Geocoding program ---*/
dm 'af cat=SASHELP.GIS.GEOCODEB.SCL';

/*--- Show geocoding results on a bar chart -------------------*/
axis1     label=(height=1.3 'Address Status');
axis2     label=(angle=-90 rotate=90 height=1.3 'Percent');
title1    "Geocoding Results";
title2    "Wake County, NC";
footnote1 j=l "Geocoded by SAS/GIS";
proc gchart data=WORK.ADDRESSES;  /* Geocoded data set */
   vbar _status_ /                 /* Midpoint (x-axis) variable */
        descending                 /* Order of results */
        type    = pct              /* Response (y-axis) variable */
        outside = pct              /* Label on top of bars */
        inside  = freq             /* Label inside of bars */
```

```
          maxis    = axis1         /* x-axis */
          raxis    = axis2;        /* y-axis */
   run;
quit;

/*--- Set up Batch Import variables ---*/
%let imp_type = GENPOINT;       /* Importing data as points */
%let maplib   = WORK;           /* Map library */
%let mapcat   = WAKE;           /* Map catalog */
%let mapname  = TRACT;          /* Map catalog entry */
%let spalib   = WORK;           /* Spatial data library */
%let spaname  = WAKE;           /* Spatial entry name */
%let cathow   = UPDATE;         /* Append existing entry */
%let spahow   = APPEND;         /* Append to spatial data sets */
%let nidvars  = 0;              /* Put points in one layer */
%let infile   = WORK.ADDRESSES; /* Data set to import */

/*--- Run the Batch Import program ---*/
DM 'af cat=SASHELP.GISIMP.BATCH.SCL';

/*--- Modify imported layer and map with GIS Procedure ---*/
proc gis cat=WORK.WAKE;
   /*--- Set display parmameters for imported point layer */
   layer update ADDRESSES /                    /* Geocoded layer */
      type    = point                          /* Layer type */
      where   = 'node=1'                       /* Layer definition */
      des     = 'Geocoded addresses'           /* Label for entry */
      default = (point=(color     = yellow /* Symbol color */
                        font      = marker /* Symbol font */
                        character = 'V'    /* Symbol to use */
                        size      = 10));  /* Symbol height */
   /*--- Set display parmameters for the map */
   map update TRACT /                          /* Map entry name */
      layerson = (TRACT ADDRESSES)             /* Turn on layers */
      cback    = gray                          /* Background color */
      legend   = hideall                       /* Turn off legend */
      des      = 'Wake County geocoding';      /* Label for entry */
   /*--- Add label in lower right corner of the map */
   maplabel create /
      text      = 'Geocoding by SAS/GIS' /* Label text */
      map       = WORK.WAKE.TRACT        /* Map entry */
      attach_to = window                 /* Do not pan label */
      position  = (bottom right)         /* Window position */
      color     = cxA81010;              /* Text color */
   run;
quit;

/*--- Open map in SAS/GIS ---*/
dm 'gis map=WORK.WAKE.TRACT';
```

Example Results

The geocoded latitude and longitude values are written to the WORK.ADDRESSES input data set, along with the census tract values for each found address. The match results for each geocoded address are also written to that data set. A bar chart that summarizes the results of the geocoding process is generated using the GCHART procedure in SAS/GRAPH software.

When the import is complete, the map opens in SAS/GIS. The found locations are in the map's ADDRESSES point layer.

Note: The WHERE clause for the ADDRESSES point layer is `WHERE='NODE=1'`, which displays points for all of the found addresses. You can modify the WHERE clause to show only those addresses that were matched with a higher degree of certainty—for example, `WHERE='_SCORE_>=40'`. △

Note: For a more detailed example of batch geocoding, see the article "Cheap Geocoding: SAS/GIS and TIGER Data," available in the Geocoding section of the Downloads page in the SAS Maps Online area at `http://support.sas.com`. The article is a reprint of a presentation from SUGI 30 and is also available in the proceedings for that conference. △

Hints and Tips for Batch Geocoding

- To ensure good quality and accurate geocoding results, you must use accurate data. If your map's address data is incomplete or out of date, your geocoding will not deliver the results you want.
- You can import the geocoded addresses onto a map. However, before you import the points, you must make sure that your address data set contains a variable named ID that has a unique value for each point.
- The input address data set that contains the addresses that you want to geocode should contain variables for the street address, city, state, and ZIP code (and an optional ZIP+4 code) of the addresses to be matched. The address data set can also contain a name that is associated with the address, but the name is not used in the address matching. In order for the geocoding facility to most accurately parse the addresses, follow these guidelines:
  - Use only street addresses. Post office boxes, rural routes, grid addresses, and addresses with alphanumeric characters cannot be geocoded. An address containing a post office box or a rural route address in addition to a street address should not cause a problem.
  - The street number portion of the street address should not contain non-numeric characters. For example, an address such as `501-B Kent St` will be matched to `501 Kent St.`, not to the full address containing the non-numeric character. Apartment numbers should be stored in separate variables rather than appended to the street number.
  - Use the following values for directional prefixes and suffixes, with no punctuation or spaces between letters:

    `N S E W NE NW SE SW`
  - Avoid using abbreviations that conflict with street name abbreviations. For example, do not use `St John St`. Use `Saint John St` instead. Spelling out `Saint` reduces chances for confusion.

  Note: The results from the geocoding are written back to the address data set, so you must have write access to it or make a copy you can write to. △
- You can create your own geocoding lookup data sets for specific areas of the United States by downloading and importing TIGER data from the U.S. Census Bureau. For more information, see Chapter 3, "Importing Spatial Data," on page 23.

You can also download ready-to-use geocoding lookup data sets for the entire United States from the SAS Maps Online area at **`http://support.sas.com`**. After downloading and installing these data sets, you can use them to geocode any U.S. address.

CHAPTER

7

The GIS Procedure

Overview: GIS Procedure

The GIS procedure creates and maintains the spatial databases that are used by SAS/GIS software. A SAS/GIS spatial database consists of the following elements:

- SAS data sets that contain the coordinates and identifying information for the spatial features.
- a spatial entry (a SAS catalog entry of type GISSPA) that identifies which SAS data sets contain spatial information. The spatial entry also stores the following elements:
  - composites that define how the variables in the spatial data are used
  - names of the polygonal indexes that define the boundaries of area layers for the map
  - a lattice hierarchy that defines which features in the spatial data enclose or are enclosed by other features (the relationships among the polygonal variables)
  - information about the projection method that is used for the stored spatial data

A spatial entry alternatively can contain references to two or more other spatial entries that have been merged into a single spatial database.

- a coverage entry (a SAS catalog entry of type GISCOVER) that selects a subset of the spatial data that is available for display in a map.
- one or more layer entries (SAS catalog entries of type GISLAYER) that identify features that have common characteristics and specify how they are displayed as layers in the map.
- a map entry (a SAS catalog entry of type GISMAP) that specifies which layers from a particular coverage are included in a map. The map entry also stores the following information:
  - the names of attribute data sets that are associated with the map, along with definitions of how the attribute data is linked to the spatial data
  - the name of a SAS data set that contains labels for map features
  - definitions of GIS actions that can be performed when map features are selected
  - definitions for map legends
  - values for display and projection options

Note: The task of creating new SAS/GIS spatial databases from spatial data in other formats can also be performed interactively by using the GIS Spatial Data Importing window or programmatically by using the SAS/GIS Batch Import process. △

Concepts: GIS Procedure

How GIS Procedure Statements Are Processed

The GIS procedure supports RUN-group processing. RUN-group processing enables you to invoke the procedure and then submit additional procedure statements without submitting the PROC statement again.

In other SAS procedures that do not support RUN-group processing, a RUN statement that follows a block of submitted statements terminates the procedure. With RUN-group processing, a RUN statement executes the preceding block of statements, but the procedure remains active. You can continue to submit additional statements for the active procedure without resubmitting the PROC statement. For example, the following code invokes the GIS procedure, assigns a default catalog, and identifies the current spatial entry:

```
proc gis catalog=mymaps.region;
   spatial norwest;
```

Note: The SPATIAL, CATALOG, LATTICE, COPY, MOVE, and SYNC statements are immediate statements for the GIS procedure. That is, they are always processed immediately and do not require a RUN statement (although including a RUN statement does not do any harm). △

After you invoke the GIS procedure, suppose that you also want to define composites. You can submit additional GIS procedure statements to define the composites without submitting a new PROC statement, as shown in the following example:

```
composite create state /
   class=state
```

```
      var=(left=statel,right=stater);
   composite create county /
      class=area
      var=(left=countyl,right=countyr);
   composite create lat /
      class=y var=y;
   composite create lon /
      class=x var=x;
run;
```

You can end RUN-group processing and terminate the GIS procedure by submitting a QUIT statement:

```
quit;
```

Submitting another PROC step, a DATA step, or an ENDSAS statement also ends RUN-group processing and terminates the GIS procedure.

Note: Certain error conditions might also terminate the GIS procedure. If this occurs, a message is printed in the SAS log. △

Data Set Names in the GIS Procedure

You can specify a data set by its complete two-level name as in *libref.data-set.* If you omit *libref*, the data set is assumed to be in the library specified in the CATALOG= option in the PROC GIS statement or in the catalog that was specified by the most recent CATALOG statement.

Note: If a one-level catalog name was used in the CATALOG= option or CATALOG statement, or if no default catalog has been named, the default library is WORK, for example, WORK.*data-set.* △

Catalog Entry Names in the GIS Procedure

You can specify a GIS catalog entry by its complete three-level name, *libref.catalog.entry.* If you use only the one-level *entry*, the entry is assumed to be in the catalog that is specified in the CATALOG= option in the PROC GIS statement or in the catalog specified by the most recent CATALOG statement.

Note: If the *libref* was omitted from the CATALOG= option or catalog statement, the default library is WORK. If no default catalog has been declared, and a one-level *entry* name is used, then an error is written to the log because of insufficient information to identify the entry. △

Syntax: GIS Procedure

PROC GIS <CATALOG=<*libref.*>*catalog*>;

CATALOG <*libref.*>*catalog*;

SPATIAL <*operation*> <*libref.catalog.*>*spatial-entry* </ *options*>;

COMPOSITE *operation composite-name* </ *options*>;

POLYGONAL INDEX *operation polygonal-index* </ *options*>;

LATTICE *outer-composite-name-1* ENCLOSES *inner-composite-name-1*
<*...outer-composite-name-n* ENCLOSES *inner-composite-name-n*>

<_UNIVERSE_ ENCLOSES *inner-composite-name*>;
COVERAGE *operation* <*libref.catalog.*>*coverage-entry* </ *options*>;
LAYER *operation* <*libref.catalog.*>*layer-entry* </ *options*>;
LAYERLABEL *operation* <*options*>;
MAP *operation* <*libref.catalog.*>*map-entry* </ *options*>;
MAPLABEL<*operation*> <*option*>
COPY <*libref.catalog.*>*entry*<*.type*> </ *options*> ;
MOVE <*libref.catalog.*>*entry*<*.type*> </ *options*>;
SYNC <*libref.catalog.*>*entry*<*.type*> </ *options*>;

PROC GIS Statement

Invokes the GIS procedure and specifies the default SAS catalog in which the spatial, coverage, layer, and map entries are stored.

PROC GIS <CATALOG=<*libref.*>*catalog-name*>;

Option

The CATALOG= option specifies the default SAS catalog in which the GIS spatial, coverage, layer, and map entries referred to in subsequent statements in the PROC GIS step are stored. The option can take one of the following forms:

CATALOG=<*libref.*>*catalog-name*
CAT=<*libref.*>*catalog-name*
C=<*libref.*>*catalog-name*

If the specified catalog does not already exist, it is created when a subsequent SPATIAL, COVERAGE, LAYER, or MAP statement is executed. If you omit the *libref* argument, the default SAS library, WORK, is used.

The CATALOG= argument is overridden when you perform one of the following:

- issue a CATALOG statement in conjunction with the PROC GIS statement. Subsequent statements in the GIS procedure will refer to the catalog that was named in the most recent CATALOG statement rather than to the one that is specified in the CATALOG= option in the PROC GIS statement.
- specify fully qualified (three-level) entry names in SPATIAL, COVERAGE, LAYER, or MAP statements. This temporarily overrides the default catalog for the current statement only. It does not reset any catalog that is specified with the CATALOG= option. See the descriptions of these statements for more information.

CATALOG Statement

Identifies the default SAS catalog in which GIS spatial, coverage, layer, and map entries are stored when you specify one-level catalog entry names in subsequent statements in the PROC step.

Note: The CATALOG statement permanently replaces the CATALOG= option that is specified in the PROC GIS statement. If you use the CATALOG= option in the PROC GIS statement and then submit a CATALOG statement, subsequent statements in the GIS procedure refer to the catalog that was named in the most recent CATALOG statement.

CATALOG <CONTENTS <*options*>> <*libref.*>*catalog-name*;

CONTENTS Operation

CONTENTS displays information about the entries in the specified catalog to the SAS Output window. If a catalog is not specified, CONTENTS displays the entries in the current catalog.

The <*libref.*>*catalog-name* arguments specify the SAS catalog in which the GIS spatial, coverage, layer, and map entries that are referred to in subsequent statements in the PROC GIS step are stored.

If the specified catalog does not already exist, it is created when a subsequent SPATIAL, COVERAGE, LAYER, or MAP statement is executed. If you omit *libref*, the default SAS library, WORK, is used.

You can temporarily override the CATALOG statement by specifying fully qualified (three-level) entry names in the SPATIAL, COVERAGE, LAYER, and MAP statements. This does not reset the current default catalog.

The following list contains descriptions of the CONTENTS operation options:

ET = (*entry-type-list*)
: catalog entry types (Map, Spatial Coverage, and Layer).

STATEMENT
: displays PROC GIS statements that create the specified entries.

VERBOSE
: lists all information about the catalog (type of map, layers, actions, and associated data sets).

Catalog Name Argument

The catalog name argument specifies the name of the catalog that is used in CREATE, REPLACE, and UPDATE operations in subsequent statements. The general form of the argument is

<*libref.*>*catalog-name*

If you do not specify a libref, the WORK library is used.

SPATIAL Statement

Selects the spatial entry on which subsequent statements operate, displays information about the contents of a spatial entry, creates a new spatial entry, replaces an existing spatial entry, modifies the characteristics of an existing spatial entry, or deletes a spatial entry.

SPATIAL <*operation*> <*libref.catalog.*>*spatial-entry* <*/ options*>;

Operations

Note: If you omit the *operation* keyword, the SPATIAL statement makes the specified spatial entry the current spatial entry for subsequent operations. No SPATIAL statement options can be used in a spatial assignment statement. △

CONTENTS

prints information about the specified spatial entry to the Output window, including the following:

- a list of the dependent data objects (data sets or other spatial entries) that store the spatial data
- a list of the SAS data sets (chains, nodes, details, and polygonal indexes) that store the spatial data
- a list of the composites for the spatial data
- the lattice hierarchy for the spatial data
- the storage projection characteristics of the spatial data

No additional arguments (other than the spatial entry name) are used with this operation. An error occurs if the specified spatial entry does not exist.

Note: The specified spatial entry does not become the current spatial entry for subsequent operations unless no spatial entry is currently selected. △

CREATE

generates a new spatial entry in which subsequent composites, polygonal index names, and lattice hierarchies that are specified in the GIS procedure are stored. The new spatial entry becomes the current spatial entry for subsequent operations.

An error occurs if a spatial entry with the specified name already exists. The SPATIAL CREATE statement does not overwrite existing spatial entries. Use SPATIAL REPLACE to replace an existing entry.

For a SPATIAL CREATE statement, you must also specify both the CHAINS= and NODES= arguments or the MERGE= argument.

DELETE

deletes the specified spatial entry. By default, any polygonal index data sets that are referred to in the spatial entry are also deleted. The chains, nodes, or details data sets that are referred to in the spatial entry are not deleted. To retain existing polygonal index data sets when the spatial entry is deleted, use the KEEP argument in the SPATIAL DELETE statement.

KEEP is the only additional argument (other than the spatial entry name) that can be used with this operation. An error occurs if the specified spatial entry does not exist.

Note: For the DELETE operation, you can also specify the special value `_ALL_` for the spatial entry name argument to delete all spatial entries in the current catalog. △

CAUTION:
Use the DELETE operation with care. The GIS procedure does not prompt you to verify the request before deleting the spatial entry. Be especially careful when you use the `_ALL_` keyword. △

REPLACE
overwrites the specified spatial entry or creates a new entry if an entry with the specified name does not exist. The specified spatial entry becomes the current spatial entry for subsequent operations. The SPATIAL REPLACE statement has the effect of canceling all previously issued SPATIAL CREATE, COMPOSITE, POLYGONAL INDEX, and LATTICE statements for the specified spatial entry.

For the SPATIAL REPLACE statement, you must specify both the CHAINS= and NODES= arguments or the MERGE= argument.

UPDATE
modifies the specified spatial entry by applying new values for specified arguments. The updated spatial entry becomes the current spatial entry for the subsequent operations.

An error occurs if there is no existing spatial entry with the specified name.

Spatial Entry Name Argument

The spatial entry name argument identifies the GISSPA-type entry that you want to create, replace, update, delete, or make the current spatial entry. The general form of the argument is

<*libref.catalog.*>*spatial-entry*

CAUTION:
Do not use host commands to move or rename SAS data sets that are referenced in GISSPA entries. Moving or renaming a data set that is referred to in a spatial entry breaks the association between the spatial entry and the data set. To prevent breaking the association, use the PROC GIS MOVE statement with the CHECKPARENT option instead of a host command. △

Options

When you specify CREATE, REPLACE, or UPDATE for the *operation* keyword, you can specify one or more of the following optional arguments after the spatial entry name. When you specify DELETE for the *operation* keyword, only the KEEP option is allowed.

Note: Separate the list of arguments from the spatial entry name with a slash (/). △

CARTESIAN | LATLON
specifies the coordinate system that is used in the stored spatial data.

CARTESIAN
data is in an arbitrary rectangular (plane) coordinate system.

LATLON
data is in a geographic (spherical) coordinate system.
The default is LATLON.

Note: The CARTESIAN and LATLON arguments are ignored when the MERGE= argument is used. △

CHAINS=*data-set*
names the SAS data set that contains chain definitions for the spatial database. A chain is one or more line segments that connect one node (or point on the map) to another. For example, a series of chains can represent a railroad or a river.

Note: The CHAINS= argument is required when you use the CREATE or REPLACE keyword and do not specify the MERGE= argument. △

DEGREES | RADIANS | SECONDS

specifies the coordinate units for the stored spatial data when the coordinate system is geographic (LATLON). The default is RADIANS.

Note: This argument is ignored when the CARTESIAN or MERGE= arguments are used. △

DESCRIPTION='*string*'

specifies a descriptive phrase, up to 256 characters long, that is stored in the description field of the spatial entry. The default description is blank.

DETAILS=*data-set*

names the SAS data set that contains detail definitions for the spatial database. The endpoints of a chain are nodes. Details are the intermediate points along a chain between the nodes that delineate angle breaks in chains. They provide a finer granularity for the chain's line segments. A data set that contains detail definitions might describe the curvy outline of a coastal road.

EAST | WEST

specifies the hemisphere in which the spatial data points lie. The default is EAST. EAST refers to points east of the Prime Meridian (0 degrees longitude) at Greenwich, England, while WEST refers to points west of the Prime Meridian.

If your data is in the Western Hemisphere, longitude values (the X coordinates) are negative, that is -35° 45' 08". If your data is in the Western Hemisphere but has positive longitudes, your map is displayed flipped or with the east and west directions reversed. See Chapter 2, "Preparing Spatial Data," on page 13 for an example of this behavior. Applying the WEST argument to the spatial data causes the longitudes to be negated when the data is read in, and the map is displayed correctly.

Note: This argument is ignored when the CARTESIAN or MERGE= arguments are used. △

KEEP

specifies that polygonal index data sets are not deleted when the spatial entry is deleted. This option is valid only with the DELETE operation.

MERGE=(*spatial-entry-list*) <EDGEMATCH <LINKONLY> | OVERLAP <ZEROMISS>> <ERROR_ROW=*integer*>

builds a new spatial entry by referencing two or more existing spatial entries. The dependent data sets for the spatial entries are not actually combined when you use the MERGE argument; the new spatial entry includes them by reference. An error occurs if any of the specified spatial entries do not exist.

You can specify any of the following additional arguments in conjunction with the MERGE= argument:

EDGEMATCH <LINKONLY>

matches common boundaries between the merged spatial entries. Missing values along common boundary chains are filled in where possible by using values from the adjoining spatial data sets. The affected chains data sets are rewritten unless the LINKONLY option is specified, and you cannot reverse the operation.

ZEROMISSING

treats any left/right attribute value of zero as a missing value. Otherwise zero is considered to be a valid value when performing an EDGEMATCH merge.

ERROR_ROW=*integer*

prints an enhanced error message for the specified spatial data row during an EDGEMATCH merge operation. This option can be useful for determining what caused a specific row to fail to merge. The basic log warning will print the row number for unmatched chains in all of the merged data sets. Any of these chain numbers can be used as the ERRORROW= target.

OVERLAP
merges spatial entries without attempting to match boundaries. The chains data sets for the merged entries are not rewritten. This is the default behavior.

MULT=*multiplier-value*
specifies a constant value by which the stored spatial data coordinates are multiplied. The default is MULT=1.

Note: This argument is ignored when the MERGE= argument is used. △

NODES=*data-set*
names the SAS data set that contains node definitions for the spatial database. Nodes are the endpoints of map chains. A node can also be a single map feature represented by a point. A single node can be the endpoint for multiple chains, as at a street intersection.

Note: The NODES= argument is required when you use the CREATE or REPLACE keyword and do not specify the MERGE= argument. △

NORTH | SOUTH
indicates the hemisphere in which the spatial data points lie. The default is NORTH.

If your data is in the southern hemisphere (below the equator), latitude values (the Y coordinates) are negative, for example, -45° 12' 33". If your data is in the southern hemisphere, but the latitude values are positive, your map is displayed inverted with the north and south directions reversed. Applying the SOUTH argument to the spatial data causes the latitude values to be negated when the data is read in, and the map is displayed with the correct side up.

Note: This argument is ignored when the CARTESIAN or MERGE= arguments are used. △

Details

A spatial entry is a SAS catalog entry of type GISSPA that defines the components of a SAS/GIS spatial database. The definition specifies which SAS data sets contain spatial information, how the data sets are related, and what roles the variables play.

Any composites, polygonal indexes, and lattice hierarchies that are created or updated during an invocation of the GIS procedure are stored in the current spatial entry. Any subsequent COVERAGE statements that are issued within the PROC GIS step subset the data in the current spatial entry.

No additional arguments (other than the spatial entry name) are used when the *operation* keyword is omitted. An error occurs if there is no existing spatial entry that has the specified name.

Note: When creating or replacing spatial entries, you can either define entirely new spatial entries or merge two or more existing spatial entries. △

Examples

Define the Current Spatial Entry

The following code fragment makes MAPS.NC.NC.GISSPA the current spatial entry that is used for subsequent operations:

```
proc gis cat=maps.nc;
   spatial nc;
```

Update an Existing Spatial Entry

The following code fragment replaces the existing details data set with MAPS.USAD for the existing MAPS.USA.USA.GISSPA spatial entry:

```
spatial update maps.usa.usa / details=maps.usad;
```

Merge Three Existing Spatial Databases

The following code fragment creates a new spatial entry that is named TRIANGLE.GISSPA in the current catalog by merging three existing spatial entries, ORANGE, DURHAM, and WAKE. In this example, each of the spatial entries to be merged is stored in a different library. See Chapter 5, "Working with Spatial Data," on page 55 for more information about merging.

```
spatial create triangle / merge=(gmap1.orange.orange,
                                 gmap2.durham.durham,
                                 gmap3.wake.wake);
```

COMPOSITE Statement

Defines, modifies, or deletes associations between variables in the chains and nodes data sets.

COMPOSITE *operation composite-name </ options>*;

Operations

CREATE

defines associations between variables in the chains and nodes data sets and stores these composites in the current spatial entry.

A warning is issued and processing of the current RUN group is halted if a composite with the specified name already exists. The COMPOSITE CREATE statement does not overwrite existing composites. Use COMPOSITE REPLACE to overwrite an existing composite.

Note: Not all spatial database variables are composites of multiple SAS data set variables. Some composites represent a single SAS data set variable. △

DELETE

deletes the specified composite from the current spatial entry.

No additional arguments (other than the composite name) are used with this operation. A warning is issued and processing of the current RUN group is halted if the specified composite does not exist.

Note: The DELETE operation of the COMPOSITE statement removes a composite from the spatial entry but does not delete the SAS variables from their respective SAS data sets. △

For the DELETE operation, you can also specify the following alternative forms for the *composite-name* argument:

- □ a list of composite names, separated by spaces, to delete more than one composite in a single DELETE operation
- □ the special value **`_ALL_`** to delete all the composites in the current spatial entry

CAUTION:

Use DELETE with care. The GIS procedure does not prompt you to verify the request before deleting an existing composite. Be especially careful when you use **`_ALL_`**. △

REPLACE

overwrites the previous definition of a composite in the current spatial entry, or creates a new composite if the specified *composite-name* did not previously exist.

UPDATE

applies new values for the specified arguments to an existing composite.

A warning is issued and processing of the current RUN group is halted if there is no existing composite with the specified name.

Composite Name Argument

The composite name argument names the composite that you want to create, replace, delete, or update.

The *composite-name* value must conform to the rules for SAS names:

- □ The name can be no more than 32 characters long.
- □ The first character must be a letter or underscore (_). Subsequent characters can be letters, numeric digits, or underscores. Blanks are not permitted.
- □ Mixed-case names are honored for presentation purposes. However, because any comparison of names is not case-sensitive, you cannot have two names that differ only in case (for example, State and STATE are read as the same name).

Options

When you specify CREATE, REPLACE, or UPDATE for *operation* in a COMPOSITE statement, you can specify one or more of these options to follow the *composite-name*.

Note: Separate the list of options from the *composite-name* with a slash (/). △

BILATERAL

indicates that the composite is a left/right type for spatial data variables that apply to the left and right sides of chains. BILATERAL composites are used to define polygonal layers in a LAYER statement by denoting chains that have different left and right values. This argument provides an implicit VAR= argument, where the LEFT= and RIGHT= variable names are constructed by appending **`L`** and **`R`** to the specified composite name. For example, the following two statements are equivalent:

```
composite create state / class=area bilateral;
composite create state / class=area
                         var=(left=statel,right=stater);
```

CLASS=*class-type*

defines the role of the composite in the spatial database. The CLASS= option links specific functionality to particular composites. The default is CLASS=CLASSIFICATION.

The *class-type* value for the CLASS= option can be one of the following:

ADDRESS

indicates that the composite defines addresses in the chains data set that is used for geocoding.

Data set address values are the numeric portion of a street address, for example, the `100` in the street address, `100 North Main Street`. A chain has four values to define the address range for each side:

FROMLEFT	beginning address on the left side.
TOLEFT	ending address on the left side.
FROMRIGHT	beginning address on the right side.
TORIGHT	ending address on the right side.

When you use specify ADDRESS for the *class-type* value, you must use the following form of the VAR= argument:

VAR=(<FROMLEFT=>*variable*, <FROMRIGHT=>*variable*,
<TOLEFT=>*variable*, <TORIGHT=>*variable*)

AREA

indicates that the composite defines polygonal areas.

For polygonal areas that represent political subdivisions, you can specify the following alternative *class-type* values to indicate which features the areas represent:

COUNTRY

indicates that the composite defines countries in the chains data.

COUNTY

indicates that the composite defines counties in the chains data.

STATE

indicates that the composite defines states in the chains data. Composites of this class are used in geocoding.

When you use AREA (or COUNTRY, STATE, or COUNTY) for the *class-type* value, you must specify the bilateral form of the VAR= argument to specify the variables that identify the features on the left and right sides of each chain in the area:

VAR=(<LEFT=>*variable*, <RIGHT=>*variable*)

CITY | PLACE

indicates that the composite defines features that are related to geographic location, such as cities. Composites of this class are used in geocoding.

By default, CITY is not considered an AREA-type composite. If your spatial data contain closed city boundaries, you must explicitly define the composite as an AREA class as well:

```
composite create towns / var=(cityl cityr) class=(city area);
```

CLASSIFICATION

indicates that the composite defines a general descriptive value that can be used to classify features in the map.

Note: In order to create new point layers when you add points to the map interactively in the GIS Map window, you must define at least one CLASSIFICATION-type composite in the spatial entry. △

DIRECTION_PREFIX
indicates that the composite defines the directional prefix component of an aggregate feature name, such as the **`North`** in **`North Main Ave`**. Composites of this class are used in geocoding.

DIRECTION_SUFFIX
indicates that the composite defines the direction suffix component of an aggregate feature name, such as the **`South`** in **`2nd St South`**. Composites of this class are used in geocoding.

NAME
indicates that the composite defines the names of features in the chains data, such as **`Central Park`**, or the name component of an aggregate feature name, such as the **`Main`** in **`E Main St`**. Composites of this class are used in geocoding.

PLUS4
indicates that the composite defines extended postal delivery codes (U.S. ZIP+4) in the chains data. Composites of this class are used in address matching.

By default, PLUS4 is not considered an AREA-type composite. If your chains data contain closed ZIP+4 boundaries, you must explicitly define the composite as an AREA class as well:

```
composite create zip4 / var=(zip4l zip4r) class=(area plus4);
```

TYPE
indicates that the composite defines the feature type component of an aggregate feature name, such as the **`Ave`** in **`N Harrison Ave`**. Composites of this class are used in geocoding.

X
indicates that the composite defines the X coordinates for the nodes in the nodes data set.

Y
indicates that the composite defines the Y coordinates for the nodes in the nodes data set.

ZIPCODE
indicates that the composite defines postal delivery codes in the chains data. Composites of this class are used in geocoding.

By default, ZIPCODE is not considered an AREA-type composite. If your chains data set contains closed ZIP code area boundaries, you must explicitly define the composite as an AREA class as well:

```
composite create zip / var=(zipl zipr)
                       class=(zipcode area);
```

VAR=*association-declaration*
defines a variable or an association between related variables in the current spatial chains or nodes data set. Variables for all composites are assumed to be in the chains data set except for CLASS=X and CLASS=Y variables, which must be in the nodes data set.

The VAR= argument is required when you use the CREATE or REPLACE operations, except in the following circumstances:

- If you omit the VAR= argument and specify CLASS=CLASSIFICATION (or omit the CLASS= argument), the *composite-name* that you specify is also used as the variable name. For example, the following statements are equivalent:

```
composite create cfcc;
composite create cfcc / var=cfcc class=classification;
```

- If you omit the VAR= argument and specify one of the bilateral *class-type* values such as AREA or STATE, the suffixes **L** and **R** are added to the *composite-name* to form the variable name pair for the association. For example, the following statements are equivalent:

```
composite create state / class=state;
composite create state / class=state
                         var=(statel stater);
```

  For other *class-type* values, the VAR= argument is required when you use the CREATE or REPLACE keywords.

The *association-declaration* argument for the VAR= option can be one of the following, depending on the *class-type* values that are specified in the CLASS= option:

variable
: declares a composite consisting of a single SAS variable. Use this form for single-variable association classes such as CLASSIFICATION, DIRECTION_PREFIX, DIRECTION_SUFFIX, NAME, TYPE, X, and Y.

(<LEFT=>*variable-1*, <RIGHT=>*variable-2*)
: declares a composite consisting of two variables that represent the left and right sides of a feature. Association declarations of this form can be used to define the boundaries between elements in the spatial data. Use this form for bilateral association classes such as AREA, CITY, COUNTRY, COUNTY, PLACE, STATE, ZIPCODE, and PLUS4.

(<FROMLEFT=>*variable-1*, <FROMRIGHT=>*variable-2*, <TOLEFT=>*variable-3*, <TORIGHT=>*variable-4*)
: declares a composite that consists of four variables that separately represent the beginning and end of the left and right sides of a feature. Association declarations of this form can be used to define the locations of specific addresses in the spatial data. Use this form for the ADDRESS class.

Note: *Variable* is the name of a SAS data set variable in the chains data set. An error occurs if any of the specified variables do not exist in the chains data set. △

Details

Once defined, composites can be referenced by other GIS procedure statements. For example, if a spatial database contains the variables COUNTYL and COUNTYR that identify the chains' left and right values for a county ID variable, you could use the COMPOSITE statement to create a composite called COUNTY by associating the two spatial database variables. The COUNTY composite could then be used to define the county boundaries for the map.

Composites are stored in the currently specified spatial (GISSPA) entry. An error occurs if you submit a COMPOSITE statement when no spatial entry is currently selected.

Note: Use the SPATIAL CONTENTS statement to view the composites for a spatial entry. Composite variable values are also displayed in the Spatial Information window when you select a map feature in the GIS Map window. △

Examples

Define a Single-Variable Composite

The following code fragment associates the class Y with the variable named LAT in the nodes data set to indicate that the variable contains north-south coordinate information:

```
composite create latitude / var=LAT class=y;
run;
```

Define a Composite for a Bilateral Feature

Both of the following code fragments associates a pair of variables in the chains data set that contain values for the left and right sides of area boundaries:

```
composite create state / var=(left=statel,right=stater)
                          class=area;
run;

composite create state/ bilateral
                         class=area;
run;
```

Define a Composite for an Address Feature

The following code fragment associates two pairs of variables in the chains data set that contain values for the corners of address boundaries:

```
composite create custadd /
   var=(fromleft=FRADDL,fromright=FRADDR,
        toleft=TOADDL,toright=TOADDR)
   class=address;
run;
```

POLYGONAL INDEX Statement

Creates, replaces, modifies, or deletes polygonal index data sets by using a libref and polygonal index references from a spatial entry.

POLYGONAL INDEX *operation polygonal-index* <*/ options*>;

Operations

CREATE

creates a polygonal index data set and stores the polygonal index definition in the current spatial entry.

A warning is issued and processing of the current RUN group is halted if either a polygonal index definition or a SAS data set with the specified names already exist. The POLYGONAL INDEX CREATE statement does not overwrite existing index definitions or data sets. Use POLYGONAL INDEX REPLACE to replace an existing index definition or data set.

For a POLYGONAL INDEX CREATE statement, you must specify both the COMPOSITE= and OUT= arguments.

DELETE
removes the specified polygonal index definition from the spatial entry. By default, the POLYGONAL INDEX DELETE statement also deletes the associated index data set. You can use the KEEP option to prevent the index data set from being deleted.

KEEP is the only additional argument (other than the polygonal index name) that can be used with this operation. A warning is issued and processing of the current RUN group is halted if the specified polygonal index does not exist.

For DELETE, you can also specify the special value `_ALL_` for the *polygonal-index* argument to delete all the polygonal index definitions in the current spatial entry.

CAUTION:
Use DELETE with care. The GIS procedure does not prompt you to verify the request before deleting an existing polygonal index. Be especially careful when you use the `_ALL_`. △

REPLACE
overwrites the polygonal index definition in the current spatial entry or creates a new polygonal index definition if the specified index does not exist.

For a POLYGONAL INDEX REPLACE statement, you must specify both the COMPOSITE= and OUT= arguments.

Note: If the data set that is specified in the OUT= argument already exists and belongs to a different spatial entry, you must specify the FORCE argument to cause it to be overwritten. △

UPDATE
modifies only the specified characteristics for an existing polygonal index.

A warning is issued and processing of the current RUN group is halted if there is no existing polygonal index with the specified name.

Note: If the data set that is specified in the OUT= argument already exists and belongs to a different spatial entry, you must specify the FORCE argument to cause it to be overwritten. △

Polygonal Index Name Argument

The polygonal index name argument names the polygonal index you want to create, delete, replace, or update.

The *polygonal-index* value must conform to the rules for SAS names:

- The name can be no more than 32 characters long.
- The first character must be a letter or underscore (_). Subsequent characters can be letters, numeric digits, or underscores. Blanks are not permitted.
- Mixed-case names are honored for presentation purposes. However, because any comparison of names is not case-sensitive, you cannot have two names that differ only in case (for example, State and STATE are read as the same name).

Options

When you specify CREATE, REPLACE, or UPDATE for the *operation* in a POLYGONAL INDEX statement, you can specify the following additional arguments following the polygonal index name. When you specify DELETE for the *operation* keyword, only the KEEP option is allowed.

Note: Separate the list of arguments from the polygonal index name with a slash (/). △

AREA

calculates the enclosed areas and perimeter lengths for the lowest-level area composite that is specified on the COMPOSITE= argument. The calculated area is added to the polygonal index data set in a variable named AREA. A label for the AREA variable contains the storage area units. The calculated perimeter is added to the polygonal index data set in a variable named PERIMETER. A label for the PERIMETER variable contains the units.

CENTROID < = GEOMETRIC | VISUAL >

CENTROID=GEOMETRIC

returns the actual calculated centroids, which might not fall within the boundaries of their corresponding polygons. The coordinates are added to the polygonal index data set in variables that are named CTRX and CTRY. Labels for the CTRX and CTRY variables contain the storage projection units and indicate that this is a GEOMETRIC centroid. Specifying the CENTROID argument by itself returns the same results as specifying CENTROID=GEOMETRIC.

CENTROID=VISUAL

returns adjusted centroids that are moved to be within the boundaries of their corresponding polygons. The coordinates are added to the polygonal index data set in variables that are named CTRX and CTRY. Labels on the CTRX and CTRY variables contain the storage projection units and indicate that this is a VISUAL centroid.

COMPOSITE=(*composite-name-1*<, ..., *composite-name-n*>)

specifies the composite or list of composites that define the boundaries of the enclosed polygonal areas that are used to create the index. If the *composite-name* list consists of a single composite, you can omit the parentheses. An error occurs if any of the specified composites are not defined in the current spatial entry or if any do not have the CLASS attribute of AREA.

Note: The COMPOSITE= argument is required when you use the CREATE or REPLACE operation. △

ERRORS<=*number*>

specifies whether messages about any topological errors that are detected while the index is being constructed are written to the SAS log. A polygon boundary consists of a single chain with the same starting and ending node, or multiple chains that form a closed boundary. The starting node for each boundary chain must be the ending node of the previous chain. The ending node of the last chain must be the beginning node of the first boundary chain. A topology error occurs when the polygon is not closed. You can specify the ERRORS argument with no added parameter to print all topological error messages, or you can add the *=number* parameter to specify the maximum number of topological error messages that will be written to the log.

FORCE

indicates that an existing polygonal index data set that is specified in the OUT= argument can be overwritten, even if it belongs to a different spatial entry. If you omit this option, the data set is not replaced and a warning is issued.

KEEP

specifies that polygonal index data sets are to be retained when the index definition is removed from the spatial entry. This option is valid only with the DELETE operation.

OUT=*data-set-name*
names the index data set that you want to create, replace, or update.

Note: The OUT= argument is required when you use the CREATE or REPLACE operation. △

CAUTION:
Do not use host commands to move or rename polygonal index data sets. Because the polygonal index data set names are stored in GISSPA entries, moving or renaming a polygonal index data set breaks the association between the GISSPA entry and the data set. To prevent breaking the association, use the PROC GIS MOVE statement with the CHECKPARENT option instead of a host command. △

Details

Polygonal indexes delineate enclosed areas in the spatial data by noting the chains that form polygons. This statement is also used to compute the enclosed areas, the centroid coordinates, and the perimeter lengths of the individual polygons.

The spatial database must include a polygonal index data set for each feature type that you intend to represent as an area layer in the map. For example, to represent states and counties as enclosed areas, you must have separate polygonal indexes for each.

The POLYGONAL INDEX statement uses composite values from the current spatial entry to determine area boundaries. The composites that are used for polygonal indexes must have the CLASS attribute AREA (or one of the political subdivision area classes such as COUNTRY, STATE, or COUNTY that imply AREA by default).

Polygonal index definitions are stored in the currently specified spatial entry. An error occurs if you submit a POLYGONAL INDEX statement when no spatial entry is currently selected.

Note: You can use the SPATIAL CONTENTS statement to view the polygonal index definitions for a spatial entry. △

Example

The following code fragment builds a polygonal index data set that is named GMAPS.STATEX. The data set identifies the boundaries of the polygons for the area feature that is identified by the STATE composite in the current spatial entry:

```
polygonal index create state / composite=state
                               out=gmaps.statex;
run;
```

LATTICE Statement

Defines the relationships between areas in a spatial database.

LATTICE *operation outer-composite-name-1* ENCLOSES *inner-composite-name-1*
<*...outer-composite-name-n* ENCLOSES *inner-composite-name-n*>
<_UNIVERSE_ ENCLOSES *inner-composite-name*>;

Operations

CONTENTS
prints information about the lattice hierarchy in the current spatial entry to the Output window. No additional arguments are used with this operation.

CREATE
creates a new lattice hierarchy in the current spatial entry. An error occurs if the spatial entry already contains a lattice. Use the REPLACE operation to replace an existing lattice.

DELETE
removes the lattice from the current spatial entry. No additional arguments are used with this operation. An error occurs if the specified coverage entry does not exist.

REPLACE
overwrites the lattice in the current spatial entry or creates a new entry if one does not exist.

Arguments

The following arguments are used with the LATTICE statement:

outer-composite-name ENCLOSES *inner-composite-name*

UNIVERSE ENCLOSES *inner-composite-name*

outer-composite-name
is an area composite that geographically contains other enclosed AREA-type composites. *Outer-composite-name* must have the CLASS attribute AREA (or one of the political subdivision area classes such as COUNTRY, STATE, or COUNTY).

You can also use the special value **_UNIVERSE_** to signify that *inner-composite-name* is a single area composite that is not contained within other enclosed areas and that does not itself enclose any other areas.

ENCLOSES
is the separator between LATTICE composites. The characters -> can be used in place of ENCLOSES.

inner-composite-name
is an area composite that is geographically within the *outer-composite-name* polygonal areas. It can also be a single area that is not contained by another when preceded by **_UNIVERSE_**. *Inner-composite-name* must have the CLASS attribute AREA (or one of the political subdivision area classes such as COUNTRY, STATE, or COUNTY).

UNIVERSE
indicates that the composite is a single area that is not part of a hierarchy.

Details

The LATTICE statement defines which areas enclose other smaller areas (such as states enclose counties). When a lattice hierarchy is defined, the area composite values for new points are assigned automatically as the points are added to the map. The composite values are also reevaluated automatically when an existing point is moved to a new location. A lattice definition also makes it possible to simultaneously assign attribute values to all points in a point layer by setting area attributes in the GIS Layer window. Area attributes cannot be assigned to new points, moved points, geocoded points, or imported points unless a lattice has been defined.

The lattice definition is written to the current spatial entry. An error occurs if you submit a LATTICE statement when no spatial entry is currently selected.

Note: Because the LATTICE statement uses composites, you must include a RUN statement following a COMPOSITE statement. This ensures that the composite is created before the LATTICE statement executes and attempts to use the composite. △

The LATTICE statement checks lattice definitions for circular references. For example, a lattice definition of the following form would cause an error:

```
LATTICE A ENCLOSES B
        B ENCLOSES C
        C ENCLOSES B;
```

Examples

Single Hierarchy

For a lattice hierarchy that comprises several composites, the general form of the LATTICE statement is

LATTICE CREATE *A* ENCLOSES *B*

B ENCLOSES *C*

C ENCLOSES *D*;

Assume that the spatial database contains states that are subdivided into counties, that the counties are further subdivided into tracts, that the tracts are further subdivided into blocks, and that corresponding composites are defined for each. The following code fragment defines the lattice for the spatial database:

```
lattice create state  encloses county
        county encloses tract
        tract  encloses block;
```

Multiple Hierarchies

You can define more than one lattice hierarchy for a spatial database, for example, when the map has overlapping AREA-type composites that are not related. A single LATTICE statement is used, but the GIS procedure recognizes the break between the two hierarchies, as follows:

```
lattice create state  encloses county /* first lattice */
        county encloses tract  /* first lattice */
        tract  encloses block  /* first lattice */
        mall   encloses store; /* second unrelated lattice */
```

Single-Element Lattice

If the map has only one AREA-type composite, it is called a universe-enclosed association. Use the **_UNIVERSE_** keyword to define a lattice for a universe-enclosed association, as follows:

```
lattice create _universe_ encloses tract;
```

It is possible to have more than one set of unrelated AREA composites, for example, a spatial entry containing counties and telephone area codes. The lattice hierarchy would then be defined as:

```
lattice create _universe_ encloses AreaCode
               _universe_ encloses County;
```

COVERAGE Statement

Displays information about the contents of a coverage entry, creates a new coverage entry, replaces an existing coverage entry, modifies the characteristics of a previously created coverage entry, or deletes a coverage entry.

COVERAGE *operation <libref.catalog.>coverage-entry </ options>*;

Operations

CONTENTS
prints information about the specified coverage entry to the Output window, including the WHERE expression that defines the spatial database subset and details of the spatial database as provided by the SPATIAL CONTENTS statement.
No additional arguments (other than the coverage entry name) are used with this operation. An error occurs if the specified coverage entry does not exist.

CREATE
creates a new coverage entry.
An error occurs if a coverage entry with the specified name already exists. The CREATE operation does not overwrite existing coverage entries. Use the REPLACE operation to replace an existing entry.
For a COVERAGE CREATE statement, you must also specify the WHERE= argument.

DELETE
removes the specified coverage entry.
No additional arguments (other than the coverage entry name) are used with this operation. An error occurs if the specified coverage entry does not exist.
For the DELETE operation, you can also specify the special value **`_ALL_`** for the coverage entry name argument to delete all coverage entries in the current catalog.

CAUTION:
Use DELETE with care. The GIS procedure does not prompt you to verify the request before deleting the coverage entry. Be especially careful when you use **`_ALL_`**. △

Note: You must specify new coverages for any map entries that refer to the deleted coverage entry. △

REPLACE
overwrites the specified coverage entry or creates a new entry if an entry with the specified name does not exist. The REPLACE operation has the effect of canceling the previously issued CREATE operation for the specified coverage entry.
For a REPLACE operation, you must also specify the WHERE= argument.

UPDATE
modifies the specified coverage entry by applying new values for specified arguments.
An error occurs if there is no existing coverage entry with the specified name.

Coverage Entry Name Argument

The coverage entry name argument identifies the coverage entry that you want to create, delete, replace, or update. The *coverage-entry* name must conform to the rules for SAS names:

- □ The name can be no more than 32 characters long.
- □ The first character must be a letter or underscore (_). Subsequent characters can be letters, numeric digits, or underscores. Blanks are not permitted.
- □ Mixed-case names are honored for presentation purposes. However, because any comparison of names is not case-sensitive, you cannot have two names that differ only in case (for example, State and STATE are read as the same name).

Options

When you specify CREATE, REPLACE, or UPDATE for the *operation* in a COVERAGE statement, you can specify one or more of these options following the coverage entry name.

Note: Separate the list of arguments from the *coverage-entry* with a slash (/). △

DESCRIPTION='*string*'
specifies a descriptive phrase, up to 256 characters long, that is stored in the description field of the GISCOVER entry. The default description is blank.

SPATIAL=*spatial-entry*
specifies the GISSPA-type entry to which the coverage definition refers. The default is the current spatial entry.

An error occurs if there is no existing spatial entry that has the specified name, or if you omit this argument when no spatial entry is currently selected.

WHERE=('*where-string-1* <... '*where-string-n*')
specifies a WHERE expression that subsets the chains data set to define a geographic coverage of a spatial database. *Where-string* can contain a complete valid WHERE expression of 200 characters or fewer.

To specify a WHERE expression greater than 200 characters, you must break the expression into separate quoted strings. When WHERE= is processed, the strings are concatenated, with a space between each string, and the entire expression is evaluated.

You can specify multiple WHERE expressions to delineate the coverage. If you are using multiple strings, each string does not have to contain a complete WHERE expression, but the concatenated expression must be valid.

You can use any of the variables in the chains data set in the WHERE expression, not just the coordinate variables. When the map is opened, only those chains that match the WHERE clause are read in. You can use only variables in the WHERE expression, not composites. Specify WHERE='1' to define a coverage that includes the entire spatial database.

Note: The WHERE= argument is required when you use the CREATE or REPLACE operation. △

Details

A coverage entry is a SAS catalog entry of type GISCOVER that contains information about the spatial data that is used to create a map. The entry also contains a subsetting WHERE expression to define the subset of spatial data, or coverage, of the map that you want to display.

For example, you could create a coverage entry, MYCAP, that contains geographic information for your state capital. MYCAP subsets the spatial database that is defined in the spatial entry MYSTATE, which contains geographic information that is used to create a map of your entire state.

Note: Even if you want to display the entire geographic scope of your spatial data and not a subset, you must still create a coverage entry by using WHERE='1'. △

Examples

Define a Universal Coverage

The following code fragment creates a coverage entry that is named GMAPS.USA.ALL.GISCOVER. The code defines a coverage of the entire spatial database that is defined in GMAPS.USA.USA.GISSPA:

```
proc gis cat=gmaps.usa;
   spatial usa;
   coverage create all / where='1';
run;
```

Define a Coverage Subset

Assume that the chains data set for the current spatial entry has the variables STATEL and STATER that contain FIPS state codes for the states on the left and right side of each chain. The following code fragment creates a coverage entry that is named SOUTHEAST of type GISCOVER. The code defines a coverage of only the selected states from the current spatial entry:

```
coverage create southeast /
   where=("statel in (1,12,13,28,37,45,47) |
           stater in (1,12,13,28,37,45,47)");
run;
```

LAYER Statement

Displays information about the contents of a layer entry, creates a new layer entry, replaces an existing layer entry, modifies the characteristics of an existing layer entry, or deletes a layer entry.

LAYER *operation <libref.catalog.>layer-entry </ options <theme-options>>*;

Operations

CONTENTS
displays the characteristics of the specified layer entry in the OUTPUT window, including the WHERE expression that defines the layer and lists of the layer's parameters and graphical attributes.
An error occurs if the specified layer entry does not exist.

CREATE
creates a new layer entry to define a particular set of features in the spatial database.
An error occurs if a layer entry with the specified name already exists. The LAYER CREATE statement does not overwrite existing layer entries. Use LAYER REPLACE to replace an existing entry.

For a LAYER CREATE statement, you must also specify either the COMPOSITE= argument or the WHERE= argument. (For area layers, you must use the COMPOSITE= argument.)

DELETE

removes the specified layer entry.

No additional arguments (other than the layer entry name) are used with this operation. An error occurs if the specified layer entry does not exist.

For the DELETE operation, you can also specify the special value `_ALL_` for the *layer-entry* name to delete all layer entries in the current catalog.

Note: You must specify a new layer list for any map entries that refer to the deleted layer entry. △

CAUTION:

Use DELETE with care. The GIS procedure does not prompt you to verify the request before deleting the layer entry. Be especially careful when you use `_ALL_`. △

REPLACE

overwrites the specified layer entry or creates a new layer entry if an entry with the specified name does not exist. The LAYER REPLACE statement has the effect of canceling the previously issued LAYER CREATE statement for the specified layer entry.

For a LAYER REPLACE statement, you must also specify either the COMPOSITE= argument or the WHERE= argument. (For area layers, you must use the COMPOSITE= argument.)

UPDATE

modifies the specified layer entry by applying new values for specified arguments.

An error occurs if there is no existing layer entry with the specified name.

Layer Entry Argument

The layer entry name argument identifies the layer entry that you want to create, delete, replace, or update. The general form of the argument is

<libref.catalog.>layer-entry

The *layer-name* must conform to the rules for SAS names:

- □ The name can be no more than 32 characters long.
- □ The first character must be a letter or underscore (_). Subsequent characters can be letters, numeric digits, or underscores. Blanks are not permitted.
- □ Mixed-case names are honored for presentation purposes. However, because any comparison of names is not case-sensitive, you cannot have two names that differ only in case (for example, State and STATE are read as the same name).

Options

When you specify CONTENTS, CREATE, REPLACE, or UPDATE for *operation* in a LAYER statement, you can specify one or more additional arguments after the layer entry name.

Note: Separate the list of arguments from the *layer-entry* name with a slash (/). △

COMPOSITE=*composite-name*

specifies a composite that defines the common characteristic of the features in the layer. The COMPOSITE= argument is an alternative to specifying a WHERE expression by using the WHERE= argument. For example, if you specify COMPOSITE=STATE in the LAYER statement and the composite named STATE was created with the variable association VAR=(LEFT=STATEL,RIGHT=STATER), then the implied WHERE expression for the layer is WHERE STATEL NE STATER.

Note: Either the COMPOSITE= argument or the WHERE= argument is required when you use the CREATE or REPLACE operation. For area layers, you must use the COMPOSITE= argument. △

DEFAULT=(*static-arguments*)
defines the static appearance of a layer. The following are the options:

POINT=(*point-arguments*)
defines the static appearance of the symbols in a point layer. This option enables you to specify the color, size, font, and a specific character to be used for the symbols. It is valid only when TYPE=POINT is in the layer definition. The following illustrates the syntax of POINT=:

```
POINT = (COLOR = color-name | color-code | CURRENT
         SIZE = [1..21]
         FONT=font-name
         CHARACTER='character' )
```

COLOR=
: specifies the color of the point symbol. COLOR= must specify a predefined SAS color name, an RGB color code in the form CX*rrggbb*, an HLS color code in the form H*hhhllss*, or a gray-scale color code in the form GRAY*ll*. For more information about color-naming schemes, see “SAS/GRAPH Colors” in *SAS/GRAPH: Reference*. The default color is BLACK. Specify CURRENT when you want to specify BLENDCOLORS and use this range level color as one of the colors between which to interpolate.

SIZE=
: specifies the size of the point symbol. SIZE= must specify an integer that is greater than or equal to 1, and less than or equal to 21. The default size is 8.

FONT=
: specifies the font to use for the point symbol. FONT= must specify a valid font name. The default font is MARKER. Font verification can be overridden by using the FORCE option in the LAYER statement.

CHARACTER=
: specifies the character to use for the point symbol. CHARACTER= must specify a single character in quotes. The default character is ‘W’, which in the MARKER font represents a "dot."

LINE=
defines the static appearance of the lines in a line layer. You can specify the color, width, and style to be used for the lines. Valid only when TYPE=LINE is in the layer definition. The following illustrates the syntax of LINE=:

```
LINE = (COLOR = color-name | color-code
              WIDTH = [1..20]
              STYLE = SOLID | DASHED | DOTTED)
```

COLOR=
: specifies the color of the line. COLOR= must specify a predefined SAS color name, an RGB color code in the form CX*rrggbb*, an HLS color code in the form H*hhhllss*, or a gray-scale color code in the form GRAY*ll*. For more information

about color-naming schemes, see "SAS/GRAPH Colors" in *SAS/GRAPH: Reference*. The default color is BLACK.

WIDTH=
: specifies the width of the line. WIDTH= must specify an integer that is greater than or equal to 1, and less than or equal to 20. The default width is 1.

STYLE=
: specifies the style of the line. Valid values for STYLE= are SOLID, DASHED, or DOTTED. The default style is SOLID.

CENTERLINE=
defines the static appearance of the optional centerline in a line layer. The option allows you to specify whether a centerline is displayed as well as the color, width, and style to be used for the centerlines. It is valid only when TYPE=LINE in the layer definition. The following illustrates the syntax of CENTERLINE=:

```
DEFAULT = (CENTERLINE = (ON | OFF
                         COLOR = color-name | color-code
                         WIDTH = [1..20]
                         STYLE = SOLID | DASHED | DOTTED));
```

ON | OFF
: specifies whether the optional centerline is displayed. The default is OFF.

COLOR=
: specifies the color of the centerline. COLOR= must specify a predefined SAS color name, an RGB color code in the form CX*rrggbb*, an HLS color code in the form H*hhhllss*, or a gray-scale color code in the form GRAY*ll*. For more information about color naming schemes, see "SAS/GRAPH Colors" in *SAS/GRAPH: Reference*. The default color is BLACK.

WIDTH=
: specifies the width of the centerline. WIDTH= must specify an integer that is greater than or equal to 1 and less than or equal to 20. The default width is 1.

STYLE=
: specifies the style of the centerline. Valid values for STYLE= are SOLID, DASHED, or DOTTED. The default style is SOLID.

AREA=
defines the static appearance of the area fills in an area layer. Area allows you to specify the color and fill style as well as angle and spacing parameters for hatched and crosshatched fill styles. It is valid only when TYPE=AREA in the layer definition. The following illustrates the syntax of AREA=:

```
DEFAULT = (AREA = (COLOR = color-name | color-code
                   STYLE = EMPTY | FILLED | HATCH | CROSSHATCH
                   ANGLE = angle-value
                   SPACING = [2..10] ));
```

COLOR=
: specifies the fill color of the area. COLOR= must specify a predefined SAS color name, an RGB color code in the form CX*rrggbb*, an HLS color code in the form H*hhhllss*, or a gray-scale color code in the form GRAY*ll*. For more information about color-naming schemes, see "SAS/GRAPH Colors" in *SAS/GRAPH: Reference*. The default color is GRAY.

STYLE=
: specifies the fill style of the area. Valid values for STYLE= EMPTY, FILLED, HATCH, or CROSSHATCH. The default style is FILLED, which means the area contains one, solid color.

ANGLE=
: specifies an angle for hatched and crosshatched lines. ANGLE= must specify a real number that is greater than or equal to

zero and less than 90 (for crosshatch), or greater than or equal to zero and less than 180 (for hatch). The default angle for both the hatch and crosshatch is 0.

SPACING=
: specifies the spacing between hatched lines or crosshatched lines. SPACING= must specify an integer that is greater than or equal to 2 and less than or equal to 10. The lower the number, the less space between lines (the higher the number, the more space between lines). The default spacing is 7.

OUTLINE=
: defines the appearance of the area outlines in an area layer. The option allows you to specify the color, width, and style to be used for the outlines. It is valid only when TYPE=AREA is in the layer definition. The following illustrates the syntax of OUTLINE=:

```
DEFAULT = (OUTLINE = (ON | OFF
                      COLOR = color-name | color-code
                      WIDTH = [1..20]
                      STYLE = SOLID | DASHED | DOTTED));
```

ON | OFF
: specifies whether the area outline is displayed. The default is ON.

COLOR=
: specifies the color of the outline. COLOR= must specify a predefined SAS color name, an RGB color code in the form CX*rrggbb*, an HLS color code in the form H*hhhllss*, or a gray-scale color code in the form GRAY*ll*. For more information about color-naming schemes, see "SAS/GRAPH Colors" in *SAS/GRAPH: Reference*. The default color is BLACK.

WIDTH=
: specifies the width of the outline. WIDTH= must specify an integer that is greater than or equal to 1 and less than or equal to 20. The default width is 1.

STYLE=
: specifies the style of the outline. STYLE= must specify either SOLID, DASHED, or DOTTED. The default style is SOLID.

DESCRIPTION='*string*'
: specifies a descriptive phrase, up to 256 characters long, that is stored in the description field of the layer entry. The default description is blank.

DETAILON=*scale-value*
: specifies the scale at or below which detail coordinates are displayed, provided that detail points are available. This argument helps keep the detail level of a layer to a minimum when the map is zoomed to a large scale. By default, detail is displayed at all scales when detail is turned on.

Note: The DETAILON= argument is effective only when detail coordinates are read for the layer. The DETAILS argument controls whether detail coordinates are read. △

DETAILS | NODETAILS
: specifies whether the detail coordinates are read for this layer. The default is NODETAILS.

If you specify DETAILS to read the detail coordinates from the database, you can use the DETAILON= argument to control the scale at which the detail coordinates are actually displayed.

FORCE
allows you to create more than one theme by using the same variable from the same attribute data set.

MAP=*<libref.catalog.>map-entry*
identifies a GISMAP-type entry that provides theme information for layers that are created in SAS/GIS in Release 6.11 of SAS. This option is ignored for layers that are generated by later releases of SAS/GIS. For thematic layers, the link to the associated data set and the name of the response variable for the theme are stored in the map entry rather than in the layer entry. If you omit this argument, the LAYER CONTENTS statement is unable to provide thematic display information for layers that were created in SAS/GIS in Release 6.11 of SAS.

Note: The MAP= argument is valid only in conjunction with the CONTENTS and UPDATE operations and is the only option that is permitted with the CONTENTS operation. △

LABELON=*scale-value*
specifies the numeric scale at or below which map labels are displayed. This argument helps keep the number of items in the map window to a minimum when the map is zoomed to a large scale. By default, labels are displayed at all scales.

OFFSCALE=*scale-value*
specifies the scale at or below which the layer is hidden. By default, the layer is displayed at all zoom scales. The value specified for OFFSCALE= must be less than the value specified for ONSCALE=. The following illustrates the syntax of OFFSCALE=:

OFFSCALE=(<*layer-off-scale*>
<ON | OFF>
<*real-units/map-units*>
<METRIC | ENGLISH>
<NONE>)

layer-off-scale
sets a map scale where the layer is turned off when zoomed. The value is a real number.

ON | OFF
enables or disables the layer off-scale. If disabled, current scale settings remain intact.

METRIC
specifies KM/CM (kilometers per centimeter) as the units, and is the default setting.

ENGLISH
specifies MI/IN (miles per inch) as the units.

real-units/map-units
are other arbitrary combinations of units. Valid values are KM, M, CM, MI, FT, and IN. Real-units is typically KM, M, MI, or FT, and map-units is usually either CM or IN. Long forms of the unit names, for example KILOMETERS or INCH (singular or plural) are also acceptable.

NONE
disables the layer off-scale and removes all parameters.

ONSCALE=*scale-value*
specifies the scale at or below which the layer is displayed. When the map is zoomed to a larger scale, the layer is hidden. By default, the layers are displayed at all zoom scales. The following illustrates the syntax of ONSCALE=:

ONSCALE=(<*layer-on-scale*>
<ON | OFF>
<*real-units/map-units*>
<METRIC | ENGLISH>
<NONE>)

layer-on-scale
sets a map scale where the layer is turned on when zoomed. The value is a real number.

ON | OFF
enables or disables the layer on-scale. If disabled, current scale settings remain intact.

METRIC
specifies KM/CM (kilometers per centimeter) as the units, and is the default setting.

ENGLISH
specifies MI/IN (miles per inch) as the units.

real-units/map-units
are other arbitrary combinations of units. Valid values are KM, M, CM, MI, FT, and IN. Real-units is typically KM, M, MI, or FT, and map-units is usually either CM or IN. Long forms of the unit names, for example KILOMETERS or INCH (singular or plural), are also acceptable.

NONE
disables the layer on-scale and removes all parameters.

STATIC | THEMATIC

STATIC
turns the current theme off so that it is not displayed when the map is opened. It does not remove the theme from the layer entry. If the layer has no theme, STATIC is ignored. The default appearance of a newly created layer is STATIC. Use the LAYER statement's DEFAULT= option to modify static graphical attributes. See the description of the DEFAULT= option for more information.

THEMATIC
turns the current theme in the layer on so that it is displayed when the map is opened. If the layer has no theme, this option has no effect. Use the LAYER statement's THEME= option to create a theme in a layer. See the description of the THEME= option for more information.

THEME=(*theme-arguments*)
enables you to modify or delete existing themes or to create new themes. In the LAYER statement THEME argument, the *operation* argument can be one of the following:

CREATE
creates a new theme for the specified layer entry.

An error occurs if a theme already exists for the layer that uses the same variable in the same attribute data set, unless you also specify the FORCE option in the LAYER statement. The CREATE operation does not overwrite existing themes. Use REPLACE to replace an existing theme.

For a CREATE operation, you must also specify the LINK= and VAR= arguments.

REPLACE

overwrites the specified theme for the layer entry. The REPLACE operation has the effect of canceling the previously issued CREATE operation for the specified layer entry.

For a REPLACE operation, you must also specify both the LINK= argument and the VAR= arguments.

UPDATE

modifies the specified theme for the layer entry by applying new values for specified arguments.

An error occurs if the specified layer does not have at least one existing theme. For an UPDATE operation, you must specify a value for at least one of the arguments LINK=, VAR=, RANGE=, NLEVELS=, MAKE_CURRENT, or NOT_CURRENT.

If you do not specify LINK=, the current data set link is used. If you do not specify THEMEVAR=, the current thematic variable is used.

DELETE

removes the specified theme from the specified layer entry.

For a DELETE operation, you must specify a value for THEMEVAR=*variable-name* or POSITION=*integer*. An error occurs if you specify THEMEVAR=*variable-name* and if a theme based on *variable-name* does not exist.

CAUTION:

Use DELETE with care. The GIS procedure does not prompt you to verify the request before it deletes the layer theme. △

For more information about the optional arguments that you can specify with THEME= operations, see "Theme Options" on page 115.

TYPE=POINT | LINE | AREA

specifies the type of layer. The TYPE argument affects how the layer is displayed in a map. The default is TYPE=LINE.

POINT

The layer's features are discrete points and have no length or area associated with them. If a POINT feature has left and right attributes, the values of the attributes must be identical.

LINE

The layer's features have length, and they can have different values for their left and right attributes. However, a LINE feature can enclose an area, even though it is displayed as a line.

AREA

The layer's features have length and area associations and the layer is displayed as enclosed polygons.

Note: Each area layer must have a polygonal index for the composite that defines the area boundaries. △

UNITS=*unit-specification*

specifies the scale units for subsequent ONSCALE=, OFFSCALE=, and DETAILON= argument values. The default is UNITS=METRIC (for example, kilometers per centimeter). *Unit-specification* can be one of the following:

ENGLISH
selects nonmetric as the scale units, for example, miles per inch or feet per inch.

METRIC
selects metric as the scale units, for example, kilometers per centimeter or meters per centimeter.

real-units/map-units
selects a user-defined combination of units. Valid values for *real-units* and *map-units* are as follows:

- KM | KILOMETER | KILOMETERS
- M | METER | METERS
- CM | CENTIMETER | CENTIMETERS
- MI | MILE | MILES
- FT | FOOT | FEET
- IN | INCH | INCHES.

The value of *real-units* is typically KM, M, MI, or FT, and the value of *map-units* is usually either CM or IN.

WHERE=('*where-string-1* <...'*where-string-n*'>)
specifies a WHERE expression that subsets the chains data set to define a geographic layer of a spatial database. *Where-string* can contain a complete valid WHERE expression of 200 characters or fewer.

To specify a WHERE expression greater than 200 characters, you must break the expression into separate quoted strings. When WHERE= is processed, the strings are concatenated, with a space between each string, and the entire expression is evaluated.

If you are using multiple strings, each string does not have to contain a complete WHERE expression, but the concatenated expression must be valid.

You can use any of the variables in the chains data set in the WHERE expressions, not just the coordinate variables. However, the layer definition must not delineate a bounded geographic region, but rather a particular subset of the spatial data that is independent of the coverage. For example, a STREETS layer might apply to all the spatial data, even if streets do not exist in many areas. You can use only variables in the WHERE expression, not composites. Specify WHERE='1' to define a layer that contains all the features in the map.

Note: Either the WHERE= argument or the COMPOSITE= argument is required when you use the CREATE or REPLACE operation. For area layers, you must use the COMPOSITE= argument. If you use the WHERE= argument, the default layer type is LINE. △

Theme Options

When you specify CONTENTS, CREATE, REPLACE, or UPDATE for the *operation* keyword in a LAYER statement and specify the THEME= option, you can specify the following additional options.

AREA=(*argument*)
defines the appearance of the area fill for each level of the specified theme in an area layer. AREA= allows you to specify the color and fill style as well as angle and spacing parameters for hatched and crosshatched fill styles. It is valid only with TYPE=AREA in the layer definition. The following illustrates the syntax for AREA=:

```
THEME = (AREA = ((LEVEL = integer | FIRST | LAST
                  COLOR = color | CURRENT | color-code
```

```
         STYLE = EMPTY | FILLED | HATCH | CROSSHATCH
         ANGLE = angle-value
         SPACING = [2..10]) | CURRENT
        BLENDCOLOR
        BLENDSPACING));
```

ANGLE=
: specifies an angle for hatched and crosshatched lines. ANGLE= must specify a real number that is greater than or equal to zero and less than 90 (for crosshatch), or greater than or equal to 0 and less than 180 (for hatch). The default is the angle of the static area for this layer.

COLOR=
: specifies the fill color of the area. COLOR= must specify a predefined SAS color name, an RGB color code in the form CX*rrggbb*, an HLS color code in the form H*hhhllss*, or a gray-scale color code in the form GRAY*ll*. For more information about color-naming schemes, see "SAS/GRAPH Colors" in *SAS/GRAPH: Reference*. The default is GRAY. Specify CURRENT when you want to specify BLENDCOLORS and use this range level color as one of the colors between which to interpolate.

BLENDCOLOR
: interpolates the color values for any theme range levels between those specified with LEVEL=. If you want to blend between existing colors, indicate the colors with COLOR=CURRENT.

BLENDSPACING
: interpolates the hatched or crosshatched style for any theme range levels between those specified with LEVEL=. To blend between existing spacing values, indicate them as SPACING=CURRENT. If any intermediate range levels are not hatched or crosshatched, BLENDSPACING ignores them.

LEVEL=
: specifies which level of the theme is being modified. For example, LEVEL=1 refers to the first range level in this theme. LEVEL=FIRST and LEVEL=LAST can also be used to denote the initial and final range levels. If the LEVEL= arguments are omitted, the entered theme parameters are assigned to the range levels in sequence.

SPACING=
: specifies the spacing between hatched lines or crosshatched lines. SPACING= must specify an integer that is greater than or equal to 2 and less than or equal to 10. The lower the number, the less space between lines (the higher the number, the more space between lines). The default is the spacing of the static area for this layer. CURRENT is specified when you want to specify BLENDSPACING and use this range as one of the spacing values between which to interpolate.

STYLE=
: specifies the fill style of the area. Valid values for STYLE= are EMPTY, FILLED, HATCH, or CROSSHATCH. The default is the style of the static area for this layer.

CENTERLINE=(*argument*)

defines the appearance of the optional centerline for the specified theme in a line layer. The option allows you to specify whether a centerline is displayed as well as the color, width, and style to be used for the centerlines. It is valid only when TYPE=LINE is in the layer definition. The following illustrates the syntax of CENTERLINE=.

Note: A centerline does not vary in a single theme. Its appearance is the same for all range levels. △

```
THEME = (CENTERLINE = (ON | OFF
                       COLOR = color | CURRENT | color-code
                       WIDTH = [1..20]
                       STYLE = SOLID | DASHED | DOTTED));
```

COLOR=
: specifies the color of the centerline. COLOR= must specify a predefined SAS color name, an RGB color code in the form CX*rrggbb*, an HLS color code in the form H*hhhllss*, or a gray-scale color code in the form GRAY*ll*. For more information about color-naming schemes, see "SAS/GRAPH Colors" in *SAS/GRAPH: Reference*. The default is the color of the static centerline for this layer. Specify CURRENT when you want to specify BLENDCOLORS and use this range level color as one of the colors between which to interpolate.

STYLE=
: specifies the style of the centerline. Valid values for STYLE= are SOLID, DASHED, or DOTTED. The default is the style of the static centerline for this layer.

ON | OFF
: specifies whether the optional centerline is displayed. The default is the same on/off status as the static centerline for this layer.

WIDTH=
: specifies the width of the centerline. WIDTH= must specify an integer that is greater than or equal to 1 and less than or equal to 20. The default is the width of the static centerline for this layer.

COMPOSITE=(*composite-name-1* <,..., *composite-name-n*>)
: lists one or more spatial composite names when you create a new key or link for a theme. If only one composite is listed, you can omit the parentheses. The composites are paired with the attribute data set variables that are named in the DATAVAR= argument. If the composite names and the data set variable names are the same, you can specify them once with either the COMPOSITE= or DATAVAR= lists, and those names will be used for both.

Note: This is not the same argument as the COMPOSITE = argument that is used to set up a WHERE expression when you create an AREA type layer. △

DATASET=<*libref.*>*data-set*
: specifies the attribute data set when you create a new key link for a theme. If you specify a one-level data set name, the default library is WORK.

DATAVAR=(*variable-1* <,..., *variable-n*>)
: lists attribute data set variables when you create a new key link for a theme. If only one variable is listed, you can omit the parentheses. These variables are paired with the spatial composites that are named in the COMPOSITE= argument. If the data set variable names and the composite names are the same, you can specify them once with either the COMPOSITE= or DATAVAR= lists, and those names will be used for both.

LINE=(*argument*)
: defines the appearance of the line for each level of the specified theme in a line layer. The option allows you to specify the color, width and style to be used for the lines. It is valid only when TYPE=LINE in the layer definition. The following illustrates the syntax of LINE=:

```
THEME = ( LINE= ((LEVEL = integer | FIRST | LAST
                  COLOR = color | CURRENT | color-code
```

```
WIDTH = [1..20] | CURRENT
STYLE = SOLID | DASHED | DOTTED)
BLENDCOLOR
BLENDWIDTH));
```

BLENDCOLOR
: interpolates the color values for any theme range levels between those specified with LEVEL=. If you want to blend between existing colors, indicate the colors with COLOR=CURRENT.

BLENDWIDTH
: interpolates the line width for any theme range levels between those specified with LEVEL=. To blend between existing widths, indicate the widths as WIDTH=CURRENT.

COLOR=
: specifies the color of the line. COLOR= must specify a predefined SAS color name, an RGB color code in the form CX*rrggbb*, an HLS color code in the form H*hhhllss*, or a gray-scale color code in the form GRAY*ll*. For more information about color-naming schemes, see "SAS/GRAPH Colors" in *SAS/GRAPH: Reference*. CURRENT is used when you want to BLENDCOLORS and use this range level color as one of the colors between which to interpolate.

LEVEL=
: specifies which level of the theme is being modified. For example, LEVEL=1 refers to the first range level in this theme. LEVEL=FIRST and LEVEL=LAST can also be used to denote the initial and final range levels. If the LEVEL= arguments are omitted, the entered theme parameters are assigned to the range levels in sequence.

STYLE=
: specifies the style of the line. Valid values for STYLE= are SOLID, DASHED, or DOTTED. The default is the style of the static line for this layer.

WIDTH=
: specifies the width of the line. WIDTH= must specify an integer that is greater than or equal to 1 and less than or equal to 20. The default is the width of the static line for this layer. CURRENT is used when you want to specify BLENDWIDTH and use this existing range level width as one of those between which to interpolate.

LINK=*link-name*
: specifies the attribute data set containing the theme variable to be used. If you do not specify *link-name* and you are performing an update, the current data set link is used.

MAKE_CURRENT | NOT_CURRENT

MAKE_CURRENT
: specifies that the specified theme is to be the current theme when the map opens. MAKE_CURRENT is the default when a theme is created or updated.

NOT_CURRENT
: specifies that the specified theme should be created or modified but is not to be made the current theme.

NLEVELS=*integer*
: specifies the number of range levels in the theme. The value for NLEVELS must be an integer greater than one. You cannot specify both NLEVELS and RANGE=DEFAULT or RANGE=DISCRETE. If you specify NLEVELS, RANGE=*LEVELS* is assumed and can be omitted.

OUTLINE=(*argument*)
defines the appearance of the polygon outlines for each level of the specified theme in an area layer. OUTLINE= enables you to specify the color, width, and style to be used for the outlines. It is valid only when TYPE=AREA in the layer definition. The following illustrates the syntax of OUTLINE=:

```
THEME = (OUTLINE = (ON | OFF
                    COLOR = color | color-code
                    WIDTH = [1..20]
                    STYLE = SOLID | DASHED | DOTTED));
```

COLOR=
specifies the color of the outline. COLOR= must specify a predefined SAS color name, an RGB color code in the form CX*rrggbb*, an HLS color code in the form H*hhhllss*, or a gray-scale color code in the form GRAY*ll*. For more information about color-naming schemes, see "SAS/GRAPH Colors" in *SAS/GRAPH: Reference*. The default is the color of the static outline for this layer.

ON | OFF
specifies whether the area outline is displayed. The default is the same on/off status as the static outline for this layer.

STYLE=
specifies the style of the outline. Valid values for STYLE= are SOLID, DASHED, or DOTTED. The default is the style of the static outline for this layer.

WIDTH=
specifies the width of the outline. WIDTH= must specify an integer that is greater than or equal to 1 and less than or equal to 20. The default is the width of the static outline for this layer.

POINT=(*argument*)
defines the appearance of the symbol for each level of the specified theme in a point layer. The option enables you to specify the color, size, font and specific character to be used for the symbols. It is valid only when TYPE=POINT is specified in the layer definition. The following illustrates the syntax of POINT=:

```
THEME = (POINT = ((LEVEL = integer | FIRST | LAST
                   COLOR = color | CURRENT | color-code
                   SIZE = [1..21] | CURRENT
                   FONT = font1
                   CHARACTER='char')
                  BLENDCOLOR
                  BLENDSIZE));
```

BLENDCOLOR
interpolates the color values for any theme range levels between those that you specified with LEVEL=. If you want to blend between existing colors, indicate the colors with COLOR=CURRENT.

BLENDSIZE
interpolates the point size for any theme range levels between those that you specified with LEVEL=. To blend between existing sizes, indicate the sizes as SIZE=CURRENT.

CHARACTER=
specifies the character to use for the point symbol. CHARACTER= must specify a single character in quotes. The default is the character of the static point symbol for this layer.

COLOR=
: specifies the color of the point symbol. COLOR= must specify a predefined SAS color name, an RGB color code in the form CX*rrggbb*, an HLS color code in the form H*hhhllss*, or a gray-scale color code in the form GRAY*ll*. For more information on color naming schemes, see "SAS/GRAPH Colors" in *SAS/GRAPH: Reference*. CURRENT is used when you want to BLENDCOLORS and use this range level color as one of the colors between which to interpolate.

FONT=
: specifies the font to use for the point symbol. FONT= must specify a valid font name. The default is the font of the static point symbol for this layer. Font verification can be overridden by using the FORCE option in the LAYER statement.

LEVEL=
: specifies which theme range is being modified. For example, LEVEL=1 refers to the first range level in this theme. LEVEL=FIRST and LEVEL=LAST can also be used to denote the initial and final range levels. If LEVEL=1 is omitted, the entered theme parameters are assigned to the range levels in sequence.

SIZE=
: specifies the size of the point symbol. SIZE= must specify an integer that is greater than or equal to 1 and less than or equal to 21. It defaults to the size of the static point symbol for this layer. Specify CURRENT when you want to specify BLENDSIZE and use this existing range level size as one of those points between which to interpolate.

POSITION=*integer*
: specifies the position number of the target theme, starting from position 1. Negative numbers refer to positions counted backward from the last theme of the layer. For example, `POSITION=-2` refers to the second from last theme of the layer. Zero refers to the current theme, regardless of its position in the theme list. If POSITION is omitted, the default for all operations is the last theme for the layer.

RANGE= DEFAULT | DISCRETE | LEVELS
: specifies the thematic range type.

DEFAULT
: Increments are calculated automatically using an algorithm that is based on the 1985 paper by G.R. Terrell and D. W. Scott, "Oversmoothed Nonparametric Density Estimates" in the *Journal of the American Statistical Association*, Volume 80, pages 209-214.

DISCRETE
: The range is treated as a series of discrete values instead of a continuous variable. If the variable that is specified in the THEMEVAR= argument is a character variable, only RANGE=DISCRETE is allowed.

LEVELS
: The range is divided into evenly spaced increments. You do not have to specify RANGE=LEVELS if you specify NLEVELS=*integer* instead.

If you do not specify RANGE=, DEFAULT is used for numeric variables and DISCRETE is used for character variables.

THEMEVAR=*variable-name*
: specifies the theme variable in the linked attribute data set (specified in LINK=*link-name*). If you do not specify *variable-name* and you are performing an update, the current theme variable is used.

THEMEVAR=*variable-name* also specifies the theme to delete or to make current.

Details

A layer entry is a SAS catalog entry of type GISLAYER that stores information about a layer in a map. Each layer represents a different set of features on the map, but features can be displayed in more than one layer. The layer also defines how the features are displayed. For example, you could create a layer entry named RIVERS to represent the water features in your spatial data.

Layers can be displayed as either static or thematic. When a layer is displayed as static, it has a fixed set of graphical attributes (fill colors, outline colors, and so on) for all of the features in that layer. When a layer is displayed as thematic, it uses values of a response variable in an associated attribute data set to determine the graphical attributes for the layer. Information about the theme value ranges and the attribute data is stored in the layer entry.

Examples

Define a Layer Using a Composite

If the chains data set contains pairs of variables that indicate values for the areas on the left and right sides of the chains, then you can use these variable pairs to define area layers. The following code fragment defines a composite that identifies county boundaries and uses that composite to define an area layer:

```
composite create county / var=(left=countyl,right=countyr)
                          class=area;
run;
polygonal index create county / composite=county
                                out=gmaps.cntyx;
run;
layer create county / composite=county
                      type=area;
run;
```

Note: The polygonal index must be defined for the composite in order to display this area layer in a map. △

Define a Layer Using a Category Variable

Assume that the spatial database contains a variable named CFCC that contains values that identify what each chain represents. Assume also that the values of the CFCC variable for all roads begin with the letter A (A0, A1, and so on, depending on the category of road). The following code fragment defines a line layer that consists of all features that are roads:

```
layer create roads / where='cfcc =: "A"'
                     type=line;
run;
```

Note: The colon (:) modifier to the equals operator restricts the comparison to only the first *n* characters of the variable value, where *n* is the number of characters in the comparison string. The WHERE expression tests for "where the value of CFCC begins with A." △

Create a Theme

This example creates a new theme for the SASUSER.MALL.STORES map, supplied with the SAS/GIS tutorial. The theme uses the SQFT variable in the MALLSTOR attribute data set to define the theme.

```
proc gis;
   spatial sasuser.mall.mall;
   layer update sasuser.mall.store / theme = (create
                                             themevar = sqft
                                             dataset = sasuser.mallstor
                                             datavar = store
                                             composite = store
                                             link = mallstor
                                             range = discrete
                                             pos = -1
                                             not_current );
   run;
quit;
```

Update an Existing Theme

This example uses the SQFT theme that was created in the previous example and modifies it as follows:

- changes the theme variable to RENT from the same attribute data set
- breaks the RENT values into nine theme range levels
- makes the first level blue
- makes the last level **cxff0000** (red)
- blends the colors for the intermediate range levels

```
proc gis c=sasuser.mall;
   spatial mall;
   layer update store / theme=(update
                              pos = 1
                              themevar = rent
                              range = levels
                              nlevels = 9
                              area = ((level = first color = blue    )
                                      (level = last  color = cxff0000)
                                      blendcolor));
   run;
quit;
```

LAYERLABEL Statement

Applies, modifies, or deletes labels associated with a specific layer.

LAYERLABEL *operation* <*options*>;

Operations

CONTENTS

prints label information to the Output window. If you specify

- □ LAYER=, all labels associated with the specified layer are printed.

  Note: If LAYER= is omitted, every label associated with all layers in the map are printed. △
- □ _ALL_, every label in the data set associated with a layer is printed.
- □ ROW=*integer*, only the label at that row is printed.
- □ TEXT=‘*string*’, every label whose text matches the value of ‘*string*’ is printed. The text comparison is case sensitive.

If no labels are printed, a NOTE is printed to the log.

CREATE

creates a new label or labels. Unlike CREATE operations for other PROC GIS statements, duplicate labels are allowed.

DELETE

removes the specified labels and, depending on which optional arguments are specified, possibly deletes the label data set. The only valid optional arguments for DELETE are DATASET=, MAP=, LAYER=, TEXT=, ROW=, IMAGE=, and _ALL_. Any others will be ignored.

If you specify

- □ DATASET=*data-set-name* as the only argument, the label data set is deleted.
- □ MAP=*map-entry* as the only argument, the label data set reference is removed from the map entry, and the data set is deleted. If you do not specify MAP=, and you have deleted all the rows in the label data set, you are cautioned that any maps using the deleted data set will generate a WARNING when opened.
- □ TEXT=‘*string*’, every literal label having this exact string is removed from the label data set.

  Note: Literal labels are those not associated with a specific layer. △
- □ IMAGE=, the specified image is deleted.
- □ ROW=, only the label at that data set row is deleted. ROW= and _ALL_ are not allowed together. If you use ROW= and TEXT=, the TEXT= is ignored and the label at that row is deleted.
- □ _ALL_, every label associated with any layer is deleted. _ALL_ and LAYER= cannot be used together.
- □ LAYER=, every label associated with this layer is deleted.

Either DATASET= or MAP= is required or no deletions can occur.

Any DELETE operation that completely empties the label data set also causes the data set to be deleted. If a data set is deleted, a NOTE is printed to the log. If the label data set is deleted, the reference to the data set within the map entry is removed.

A note is printed in the log upon completion of a successful deletion.

REPLACE

replaces the labels for the specified layer or the specified label.

If you specify

- □ LAYER=, the labels associated with that layer are replaced. If the specified layer has no labels, a CREATE is performed.

- TEXT='*string*', the existing literal label with that string is replaced. If no label exists, a CREATE is performed.

UPDATE
: updates the labels for the specified layer or the specified label.

If you specify

- LAYER=, the UPDATE operation is limited to that layer's labels only. If the layer you specify has no labels, an ERROR is printed.
- TEXT='*string*', the existing literal label you specify is modified. If you do not specify TEXT='*string*', an ERROR is printed.

Options

When you specify CONTENTS, CREATE, REPLACE, or UPDATE for *operation* in a LAYERLABEL statement, you can specify one or more additional arguments after the layer entry name.

ALL
: affects the CONTENTS and DELETE operations as follows:.

In a CONTENTS operation, _ALL_ prints every label associated with a layer to the Output window.

In a DELETE operation, _ALL_ deletes every label associated with a layer.

ALL has no effect on CREATE, REPLACE, or UPDATE operations. If _ALL_ is detected, it is ignored.

Note: _ALL_ cannot be used in the same statement with ROW= or TEXT= options. △

ATTRIBUTE_VARIABLE=*link.variable*
: specifies a variable in an attribute data set that supplies label text for the layer.

The *link* portion of the argument is an attribute data set that is read to get the text string for each map feature to be labeled.

For each chain in the specified layer, the row number of its attribute data in the link data set is determined. The value on that row for the specified *variable* is used for the label text. The following restrictions apply to the ATTRIBUTE_VARIABLE argument:

- MAP= is required because it contains the linked attribute data set names. The link name must already exist in the map entry.
- The specified variable must already exist in the link data set.

COLOR=*color-name* | CX*rrggbb*
: specifies the text color. The default color is BLACK.

color-name
: is a SAS color-name, for example GREEN or RED.

CX*rrggbb*
: is an RGB color, for example **CX23A76B**.

For more information about color-naming schemes, see "SAS/GRAPH Colors" in *SAS/GRAPH: Reference*.

COMPOSITE=*composite-name*
: specifies a GIS composite that references a variable in a GIS spatial data set. This option is used to create labels on features in a specific map layer.

The label for each feature in the specified layer is created by first determining the row number of each map feature to be labeled. The value of the composite's associated variable for that row is then used as the label for that feature. For example, the chain whose row number in the chains data set is 35 would be labeled with the composite variable's value from row 35. The following restrictions apply to the COMPOSITE argument:

- □ COMPOSITE cannot be used with SAS_VARIABLE or ATTRIBUTE_VARIABLE options.
- □ MAP=*map-entry* is required because the map entry contains the spatial data set names.
- □ The specified composite must already exist in the map entry.

DATASET=*<libref.>data-set-name*

specifies the label data set to which new labels are appended. If the data set does not exist, it is created.

If you specify a one-level data set name, the WORK library is assumed. If you specify both DATASET= and MAP=, and the map already references a label data set, the data set names are compared. If they are not the same and FORCE was not specified, a warning is printed, and the run group is terminated.

FONT=*font-name*

fontname

specifies the font for the label text.

The following are some examples:

```
FONT = 'Times New Roman-12pt-Roman-Bold'
FONT = 'Display Manager font'
FONT = 'Sasfont (10x15) 10pt-9.7pt-Roman-Normal'
```

DEFAULT

assigns the default font to the label. If FONT= is omitted entirely, this is assumed. If the fontname specified for the label is not found when the map is opened, the default system font is substituted and a note is printed to the log.

FORCE

replaces the existing label data set reference in a map when both DATASET= and MAP= are specified. If the map already references a label data set, its data set name is compared to the name specified with DATASET=. If they are not the same, the FORCE option causes the map's label data set reference to be overwritten and a note printed to the log. The map's original label data set is not deleted.

FRONT | BACK

FRONT

causes an image label to be drawn over the map features. This is the default for image labels.

BACK

causes an image label to be drawn beneath the map features.

These options do not apply to text labels.

IMAGE=*<libref.>catalog.entry* | *'pathname'*

specifies the location of an image to use as an image label on the map.

'pathname'

enables you to enter a host directory path to an image file. For example,

```
IMAGE='C:\My SAS Files\photo.gif
```

<libref.>catalog.entry

uses an IMAGE catalog entry for the image label. If you omit the library name from the statement, the WORK library is the default.

LAYER=<*libref.catalog*>layer-entry

specifies the name of the layer with which to associate the label. The label is displayed when this layer is turned on. The labels are also placed adjacent to the features in this layer as indicated by the POSITION= option.

LAYER= is a required argument for the CREATE, REPLACE, and UPDATE operations.

The layer entry name is determined by the following rules:

- A complete three-level name entered as `libref.catalog.layer-entry` is used as-is.
- A one-level entry name can be specified. If you previously set a default libref and catalog with a PROC GIS CATALOG statement, they are used for the layer name.
- If you specify a one-level layer name, and the default assigned by a CATALOG statement is used, SAS/GIS checks to make sure the layer name matches the libref and catalog in the MAP= option. If they do not match, a WARNING is printed and the statement is ignored.
- If no default libref and catalog are active, but the MAP= option is present, that map libref and catalog is used for the layer name. A NOTE is printed to the log.

MAP=<*libref.catalog*>map-entry

indicates the map entry to display the labels on. If you indicate a one-level name, the map entry is assumed to be in the catalog that is specified in the PROC GIS statement or in the most recently issued CATALOG statement.

If the specified map entry already references a label data set, new labels are appended to that data set.

If the map entry does not reference a label data set, you must provide a label data set name with the DATASET= option. The labels are written to that data set, and the data set is then assigned to the specified map.

MAP= is a required argument.

OFFSCALE=(<*label-off-scale*> <*real-units/map-units* | METRIC | ENGLISH>)

scale

specifies a map scale where the label is turned on or off when the map view is zoomed.

units

specifies the units for OFFSCALE.

real-units/map-units

enables you to specify various combinations of units. Valid values are KM, M, CM, MI, FT, and IN. Real-units is typically KM, M, MI, or FT, and map-units is usually either CM or IN. Long forms of the unit names, for example, KILOMETERS or INCH (singular or plural), are also acceptable.

METRIC

sets the scale units to KM/CM. This is the default if units are omitted.

ENGLISH

sets the scale units to MI/IN.

OFFSET=(<*x*> <, *y*> <,> <PIXELS>| <*x* <PIXELS>> <, *y* <PIXELS>>)

specifies the distance to shift the entire label from its default location. *x* is the number of pixels to move the label right (positive numbers) or left (negative numbers), and *y* is the number of pixels to shift the label up (positive numbers) or down (negative numbers). For example,

To set only the **X** offset, specify one value, with or without a following comma:

```
OFFSET = ( 10 pixels, )
```

To set only the **Y** offset, specify one value preceded by a comma:

```
OFFSET = ( ,-30 pixels, )
```

To set both the **X** and **Y** offsets, specify two values, with or without a comma separating them:

```
OFFSET = ( 20 pixels, 40 pixels )
```

ONSCALE=(<*label-on-scale*> <*real-units/map-units* | METRIC | ENGLISH>)

scale
specifies a map scale where the label is turned on or off when the map view is zoomed.

units
specifies the units for ONSCALE.

real-units/map-units
enables you to specify various combinations of units. Valid values are KM, M, CM, MI, FT, and IN. Real-units is typically KM, M, MI, or FT, and map-units is usually either CM or IN. Long forms of the unit names, for example, KILOMETERS or INCH (singular or plural), are also acceptable.

METRIC
sets the scale units to KM/CM. This is the default if units are omitted.

ENGLISH
sets the scale units to MI/IN.

OVERLAP | NOOVERLAP
specifies how labels are treated when they overlap.

OVERLAP	All labels you create with the option will be displayed even if they conflict with other labels.
NOOVERLAP	some of the conflicting labels will be suppressed until you zoom in more closely. This is the default setting.

POINTER

POINTER	draws a leader line from the label to its associated map feature.
NOPOINTER	places the label on the map with no leader line. NOPOINTER is the default if POINTER is not present.

POSITION= (*integer-1, ..., integer-8*)
controls where the labels are placed about the features when you are labeling multiple features associated with a layer.

The new labels are associated with the map features displayed in that layer. The labels are positioned around those features to minimize conflicts and collisions. The POSITION option allows you to specify the order in which the label positions are tried. The following illustrates the syntax of POSITION=:

POSITION=(<TOP_LEFT | TL => *integer* <,>
<TOP_CENTER | TC => *integer* <,>
<TOP_RIGHT | TR => *integer* <,>
<MIDDLE_LEFT | ML => *integer* <,>
<MIDDLE_RIGHT | MR => *integer* <,>
<BOTTOM_LEFT | BL => *integer* <,>
<BOTTOM_CENTER | BC => *integer* <,>
<BOTTOM_RIGHT | BR => *integer* <,>)

The following diagram shows all of the positions around a point (X) where a label can be placed:

TL	TC	TR
ML	X	MR
BL	BC	BR

The default position values for these locations are shown in the following diagram:

6	4	1
7	X	2
8	5	3

For example, the first attempt to place the label is made in the number 1 position, and then in the number 2 position, with the number 8 position last. If the label cannot be placed in any of these positions without a collision, and if OVERLAP is not specified, the label will not be displayed when the map is opened, and a warning will be printed to the log at that time.

The following example assigns the position values as indicated:

```
POSITION=(3 1 4 7 8 5 2 6)
```

3	1	4
7	X	8
5	2	6

The following restrictions apply to the POSITION argument:

- If the keywords, for example, TOP_LEFT, are omitted and only the integer value specified, the values are assigned in left-to-right, top-to-bottom sequence. However, all eight of the values are required.
- The OFFSET= option has no effect on this form of the POSITION option. If it is encountered, it is ignored.
- Duplicate numbers are not allowed. You cannot have two locations numbered as 5. The allowed integer values are 1-8, inclusive.
- The position values are stored in the map entry. There is no need to use them in multiple LAYERLABEL statements. The last POSITION= values specified will be the ones used to determine the locations for the point label when the map is opened.

ROW=*integer*

specifies a particular label in the data set to update, replace, delete or print. It is not valid for the CREATE operations.

SAS_VARIABLE=*variable-name*

specifies a variable in the map's chains data set that is used to create labels on features in a specific map layer.

The label for each feature in the specified layer is created by first determining the row number of each map feature to be labeled. The value of the variable for that row is then used as the label for that feature. For example, the chain whose row number in the chains data set is 35 would be labeled with the variable's value from row 35. The following restrictions apply to the SAS_VARIABLE argument:

- SAS_VARIABLE cannot be used with COMPOSITE or ATTRIBUTE_VARIABLE options.
- MAP=*map-entry* is required because the map entry contains the spatial data set names.
- The specified variable must already exist in the chains data set.

TEXT=‘*string*’
specifies the text for a literal label, that is, one not associated with a specific map layer.

For REPLACE, UPDATE, DELETE or CONTENTS operations, *string* specifies a search string used to locate a specific target label if ROW= is not present. When TEXT=‘*string*’ is used to search for a label, *string* is case sensitive, and an exact match to the value of the search string must be found. For example, TEXT=‘Paris’ will ignore a label having PARIS as its text. It will also ignore a label having ‘Paris Metro’ as its text.

If ROW= and TEXT= are both used in a REPLACE or UPDATE statement, the ‘*string*’ entered with TEXT is not a search string. It becomes a replacement string for the label at the specified ROW number.

TRANSPARENT | NOTRANSPARENT

TRANSPARENT
enables the map features to show through the label's bounding box. This is the default if not specified.

NOTRANSPARENT
blocks the display of map features within the label's rectangular bounding box.

These options apply to text labels and image labels stored as catalog entries. They have no effect on image labels stored in an external file.

MAP Statement

Displays information about the contents of a map entry, creates a new map entry, replaces an existing map entry, modifies the characteristics of a previously created map entry, or deletes a map entry.

MAP *operation <libref.catalog.>map-entry </ options>*;

Operations

CONTENTS
prints information about the specified map entry to the OUTPUT window, including the following items:

- a list of the data objects (coverage and layer entries and label data set) that compose the map entry

- details of the spatial database as provided by the COVERAGE CONTENTS and SPATIAL CONTENTS statements
- details of the layer definitions as provided by the LAYER CONTENTS statement
- lists of the projection method that is used to display the map
- a list of associated data sets and link names
- a list of the GIS actions that have been defined for the map
- a list of legend definitions for the map

No additional arguments (other than the *map-entry* name) are used with this operation. An error occurs if the specified map entry does not exist.

CREATE

creates a new map entry that defines a map that can be displayed in the GIS Map window.

An error occurs if a map entry with the specified name already exists. The MAP CREATE statement does not overwrite existing map entries. Use MAP REPLACE to overwrite an existing entry.

For a MAP CREATE statement, you must also specify the COVERAGE= and LAYERS= arguments.

DELETE

removes the specified map entry.

No additional arguments (other than the map entry name) are used with this operation. An error occurs if the specified map entry does not exist.

For the DELETE operation, you can also specify the special value `_ALL_` for the map entry name argument to delete all map entries in the current catalog.

CAUTION:

Use DELETE with care. The GIS procedure does not prompt you to verify the request before deleting the map entry. Be especially careful when you use `_ALL_`. △

PRESENT

creates an HTML file to display a GIS map on the web using ODS and the IMAGEMAPS option.

REPLACE

overwrites the specified map entry or creates a new entry if an entry with the specified name does not exist. The REPLACE operation has the effect of canceling the previously issued CREATE operation for the specified map entry.

For a REPLACE operation, you must also specify the COVERAGE= and LAYERS= options.

UPDATE

modifies the specified map entry by applying new values for specified arguments.

An error occurs if there is no existing map entry with the specified name.

Map Entry Name Argument

The map entry name argument identifies the map entry that you want to create, delete, replace, or update. The general form of the argument is

<libref.catalog.>map-entry

The *map-name* must conform to the rules for SAS names:

- The name can be no more than 32 characters long.
- The first character must be a letter or underscore (_). Subsequent characters can be letters, numeric digits, or underscores. Blanks are not permitted.

- Mixed-case names are honored for presentation purposes. However, because any comparison of names is not case-sensitive, you cannot have two names that differ only in case (for example, State and STATE are read as the same name).

Options

When you specify CREATE, REPLACE, or UPDATE for the MAP operation, you can specify one or more of the following options following the *map-entry* name.

Note: Separate the list of options from the map entry name argument with a slash (/). △

ACTION=(*operation-arguments*)
The following list contains descriptions of the ACTION arguments.

COMMAND= '*command-name*' | *variable*
specifies the commands to be run when either a COMMAND or a SYSTEMCOMMAND action is executed in the map. Valid values are:

'*command-1* <;*command-2*; ...>'
To specify commands explicitly, enclose them in quotation marks. Separate multiple commands with semicolons.

variable
specifies the variable containing the commands in the linked data set.

The COMMAND= parameter is used only by the COMMAND and SYSTEMCOMMAND type actions and is a required argument. If the action type is COMMAND, COMMAND= refers to a SAS command. If the action type is SYSTEMCOMMAND, it refers to a host operating system command.

COPY
copies existing actions from one map entry to another. Specify the map entry that contains the actions to be copied with the FROM=*map-entry* argument. The actions are copied to the map that is specified in the MAP statement.

Specify the actions to be copied with the NAME=*action-name* argument. If you specify NAME=_ALL_ you copy all actions in the specified map. Existing actions in the map to be updated are not overwritten unless you specify the FORCE option in the MAP statement.

CREATE
add a new action to the map.

DELETE
removes an existing action from the map entry. Specify the action to be deleted with the NAME=*action-name* argument. You can specify NAME=_ALL_ to delete all actions. Use the NOWARN argument in the MAP statement to suppress messages when an action is not found.

CAUTION:
Use DELETE with care. The GIS procedure does not prompt you to verify the request before it deletes the action from the map. △

FORMULA= <*libref.*>*catalog.entry.type*
specifies a formula catalog entry to be used by an FSVIEW action. A FORMULA entry must be a fully qualified three- or four-level name. If the name is three levels, it is assumed to be in the WORK library. FORMULA is used only by the VIEW type action, and it is an optional argument.

FROM=*map-entry*
used with the ACTION argument COPY operation, FROM= specifies the source map entry that contains actions to be copied. Specify the actions to be copied from the map with the NAME=*action-name* argument.

IMAGEVAR=*variable-name*
specifies the name of the variable in the LINK= data set which contains the image to display for the current selected feature. IMAGEVAR is used only by an IMAGE type action, and it is a required argument.

LINK=*link-name*
specifies an attribute data set link. If the link does not exist, you can create it in the same MAP statement with the ATTRIBUTE= option. A LINK is required for all action types except a SPATIALINFO action.

MAPVAR=<*variable-name*>
specifies the name of the variable in the LINK= data set containing the three-level name of the map to be opened when a particular feature is selected. MAPVAR is used only by the TYPE=DRILLDOWN type action and is a required argument.

NAME=*action-name* | _ALL_
specifies the action to be copied, deleted, or updated. *Action-name* identifies a single action, while _ALL_ specifies all actions.

Note: You cannot specify NAME=_ALL_ if you are using ACTION UPDATE with the RENAME argument. △

OUT=*data-set-name*
specifies an output data set. OUT= is required for DATA and PROGRAM actions. It is optional for COMMAND and SYSTEMCOMMAND type actions.

OUTMODE = REPLACE | APPEND | APPEND_NEW
specifies how to the action writes to the OUTPUT data set.

REPLACE
overwrites the existing data set. REPLACE is the default.

APPEND
writes the observations to the end of the existing data set.

APPEND_NEW_
creates a new data set the first time the action is executed, and appends to this data set each additional time the action is executed.

REDISTRICT=*variable-name*
opens the Redistricting Window to adjust totals in adjoining areas.

REDISTRICTLAYER=*layer-name*
specifies the name of the polygonal layer to be themed by the redistricting action. REDISTRICTLAYER= is used only by the REDISTRICT type action and is a required argument.

REDISTRICTVAR=*variable-name*
specifies the name of the variable in the LINK data set upon which the redistricting will be based. REDISTRICTVAR= is used only by the REDISTRICT type action and is a required argument.

RENAME=*new-action-name*
renames the action that is specified in the NAME=*action-name* for UPDATE.

Note: You cannot specify RENAME if you have also specified NAME=_ALL_. △

REPLACE
replaces the named action, or, if it does not exist, creates a new action with that name.

SCREEN=<*libref.*>*catalog.entry.type*
specifies a screen catalog entry to be used by an FSBROWSE action. A SCREEN entry must be a fully qualified three- or four-level name. If the name is three levels, it is assumed to be in the WORK library. SCREEN is used only by the BROWSE type action, and it is an optional argument.

SOURCE='*filename*' | <*libref.*>catalog.*entry.type* | *fileref*
specifies the location of the source code for a PROGRAM type action. The following are valid locations:

'*filename*'
an external file containing SAS code. The host-path filename must be enclosed in quotation marks.

libref.catalog.entry.type
the three- or four-level name of a catalog entry containing the SAS code. A three-level name is assumed to be in the WORK library. Valid values for *type* are SOURCE and SCL.

fileref
a one-level name is assumed to be a SAS fileref. If the fileref does not exist, the action is created, and a warning is printed to the log.

The SOURCE parameter is used only by a PROGRAM type action and is a required argument.

SUMMARYVAR=(*variable-1* <,*variable-2*>... | _ALL_)
specifies a list of NUMERIC variables to display in the Redistricting Window when a REDISTRICT type action is executed. Only NUMERIC variables are valid because redistricting sums the values for each new district. Specifying SUMMARYVAR=(_ALL_) displays sums for every numeric variable. SUMMARYVAR is used only by the REDISTRICT type action. The default is _ALL_.

TYPE
used with CREATE to select an action type. The following are valid arguments:

BROWSE
opens an FSBROWSE window on a data set.

COMMAND
runs a SAS command.

DATA
subsets the current selections and write attribute data into a data set.

DRILLDOWN
opens another map associated with the current feature.

IMAGE
displays an image associated with the selected map feature.

PROGRAM
creates a data set and run a SAS program against its observations.

REDISTRICT
opens the Redistricting Window to adjust totals in adjoining areas.

SPATIALINFO
displays the current feature in the Spatial Info Window.

SYSTEMCOMMAND
runs a command from the host operating system.

VIEW
opens an FSVIEW window on a data set.

UPDATE
modifies existing actions in the map that is being updated. Specify the action to be updated with the NAME=*action-name* argument. You specify NAME=_ALL_ to update all actions. NAME= is required for UPDATE.

If you specify a single action, you can use the RENAME=*new-action-name* argument to change the link name. You cannot use RENAME if you specify NAME=_ALL_.

You can also change the action's execution settings with the WHEN= argument.

WHEN= OFF | IMMEDIATE | DEFERRED
used with UPDATE to change the execution setting of the specified action.

OFF
The action is not executed when a layer feature is selected.

IMMEDIATE
The action is executed as soon as a layer feature is selected.

DEFERRED
The action's execution must be performed explicitly after a layer feature has been selected.

AGGREGATE | DISAGGREGATE

AGGREGATE
sets a flag so that polygonal areas with identical ID values are considered as one. For example, if you are selecting from the STATE layer and click on North Carolina, all the Outer Banks islands are also selected.

DISAGGREGATE
sets a flag so that polygonal areas with identical ID values are treated independently. For example, if you are selecting from the STATE layer of the North Carolina map and click on Emerald Isle, only that one island gets selected. DISAGGREGATE is the default.

ATTRIBUTE=(*attribute-arguments*)
copies, deletes, or updates data links between the chains data set and attribute data sets. The following are the arguments used with ATTRIBUTE:

COMPOSITE=(*composite-name-1* <, ..., *composite-name-n*>)
lists spatial composite names when you create a new key link. These composites are paired with the attribute data set variables that are named in the DATAVAR= option. If the composite names and the data set variable names are the same, you can just specify them once with either the COMPOSITE= or DATAVAR= lists, and those names will be used for both.

COPY
copies existing attribute data links from one map entry to another. Specify the map entry that contains the links to be copied by using the FROM=*map-entry* argument. The links are copied to the map that is specified in the MAP statement.

Specify the link to be copied with the NAME=*link-name* option. If you specify NAME=_ALL_, you copy all links in the specified map. Existing links in the map to be updated are not overwritten unless you specify the FORCE option in the MAP statement.

CREATE
: adds a new attribute data link to the map.

DATASET=*libref.data-set-name*
: specifies the attribute data set when you create a new key link.

DATAVAR=(*variable-name-1* <, ...*variable-name-n*>)
: lists attribute data set variables when you create a new key link. These variables are paired with the spatial composites that are named in the COMPOSITE= option. If the data set variable names and the composite names are the same, you can just specify them once with either the COMPOSITE= or DATAVAR= lists, and those names will be used for both.

DELETE
: removes an existing attribute data link from the map entry. Specify the link to be deleted with the NAME=*link-name* argument. If you specify NAME=_ALL_, you delete all data links. Use the NOWARN option in the MAP statement to suppress messages when a link is not found. This does not delete the attribute data set, only the link.

CAUTION:
Use DELETE with care. The GIS procedure does not prompt you to verify the request before it deletes the action from the map. △

FROM=*map-entry*
: used with the ATTRIBUTE COPY operation, specifies the map entry that contains data links to be copied. Specify the links to be copied from the map with the NAME=*link-name* argument.

NAME=*link-name* | _ALL_
: specifies the attribute data link to be copied, deleted, or updated. *Link-name* identifies a single data link, while _ALL_ specifies all data links.

Note: You cannot specify NAME=_ALL_ if you are using UPDATE with the RENAME argument. △

RENAME=*new-link-name*
: renames the link that is specified in NAME=*link-name* for the UPDATE operation.

Note: You cannot specify RENAME if you have also specified NAME=_ALL_. △

UPDATE
: modifies existing data links in the map that is being updated. Specify the link to be updated with the NAME=*link-name* argument. Specify NAME=_ALL_ to update all data links. NAME= is required for the UPDATE operation.

If you specify a single link, you can use the RENAME=*new-link-name* argument to change the link name. You cannot use RENAME if you specify NAME=_ALL_.

CARTESIAN | LATLON
: specifies the coordinate system used for the displayed spatial data. The default is LATLON.

CARTESIAN
: data is in an arbitrary rectangular (plane) coordinate system

LATLON
: data is in a geographic (spherical) coordinate system.

Note: The map entry must use the same coordinate system as the spatial entry from which the map is derived. If the spatial entry specifies the CARTESIAN coordinate system, then you must also specify the CARTESIAN argument for the MAP statement. If the spatial entry specifies the LATLON coordinate system, then you must also specify the LATLON argument for the MAP statement. △

CBACK=*color*
specifies the background color of the map. CBACK= must specify a predefined SAS color name, and RGB color code in the form CX*rrggbb*, an HLS color code in the form H*hhhlllss*, or a gray-scale color code in the form GRAY*ll*. For more information about color naming schemes, see "SAS/GRAPH Colors" in *SAS/GRAPH: Reference*. The default map background color is WHITE.

COVERAGE=*<libref.catalog.>coverage-entry*
specifies the coverage entry to which the map refers. The coverage determines the geographic extent of the map.

Note: The COVERAGE= argument is required when you use the CREATE or REPLACE operation. △

DEGREES | RADIANS | SECONDS
specifies the coordinate units for the displayed spatial data when the coordinate system is geographic (LATLON). The default is RADIANS.

The unit system that you select defines the allowable range for coordinate values. For example, if you specify DEGREES, then all X coordinate values must be in the range -180 to 180, and all Y coordinate values must be in the range -90 to 90.

DEGREES	units for LATLON data are measured in decimal degrees.
RADIANS	units for LATLON data are measured in radians. RADIANS is the default.
SECONDS	units for LATLON data are measured in seconds.

DESCRIPTION='*string*'
specifies a descriptive phrase up to 256 characters long that is stored in the description field of the GISMAP entry. The default description is blank.

DETAILS | NODETAILS
specifies whether detail coordinates are read for the entire map. The default is NODETAILS.

Note: You can use the LAYER statement's DETAILS and DETAILON= options to control the display of detail coordinates for a particular layer. The MAP statement's DETAILS option overrides the LAYER statement's DETAILS option. △

FORCE
specifies that existing actions or attribute links might be overwritten during copy operations. Use this argument with the COPY argument in the ACTION or ATTRIBUTE argument.

IMAGEMAP =(HTML=(*layer-links* |)DEFAULT=*link-name*)
provides details for building an HTML version of a GIS map through ODS.

Note: The IMAGEMAP= argument is valid only with the PRESENT operation in the MAP statement. △

The PRESENT operation uses the SAS Output Delivery System (ODS) to generate an HTML page with aGIF image of the map. The GIF image can be a static image or can contain clickable map points or polygons. Each selectable map feature is associated with a URL. The URL addresses are contained in one or more variables in a SAS data set that is linked to the map.

The following options are used to specify the linked data set and the URL-related variables for specific map layers:

ALL = *variable-name*
: declares that all of the selectable map layers use the URLs stored in the specified data set variable.

HTML = (*layer-entry–1* = *variable-name* <, *layer-entry–2*=*variable-name* ...>)
: associates different URL-related variables with specific layers.

DEFAULT = *link-name*
: specifies the link name for the attribute data set that contains the URL-related variables.

LABEL= ***libref.data-set*** **| NONE | DELETE| HIDEALL | UNHIDEALL**

assigns or removes the specified label data set reference to the map. If the map already has a label data set, the original is deassigned. However, it is not overwritten.

LABEL=<*libref.*>*data-set*
: assigns the specified data set reference to the map entry. An error occurs if the specified data set does not exist. If the libref is not specified, the default WORK library is used.

LABEL=NONE
: unassigns the current label data set from the map entry, but the data set is not deleted.

LABEL=DELETE
: unassigns the current label data set from the map entry, and deletes the data set.

LABEL=HIDEALL
: hides all of the labels in the target map. HIDEALL does not remove the label data set reference from the map entry.

LABEL=UNHIDEALL
: displays all of the labels in the target map. UNHIDEALL does not display labels attached to layers that are not displayed, nor does it display labels that would not be displayed at the current scale of the map.

LAYERS=(*<libref.catalog.>layer-entry-1* <, ..., *<libref.catalog.>layer-entry-n*>)

specifies a list of GISLAYER-type entry names. The specified layers form the complete list of layers in the map entry. If the map entry already contains a list of layers, they are replaced by these layers.

Note: The LAYERS= argument is required when you use the CREATE or REPLACE operation. △

LAYERS+=(*<libref.catalog.>layer-entry-1* <, ..., *<libref.catalog.>layer-entry-n*>)

specifies a list of GISLAYER-type entry names that are added to the map's current layer list.

LAYERS-=(*<libref.catalog.>layer-entry-1* <, ..., *<libref.catalog.>layer-entry-n*>)

specifies a list of GISLAYER-type entry names that are removed from the map's current layer list. The layer entries are not deleted. They remain in their respective catalogs.

LAYERSON=(*<libref.catalog.>layer-entry-1* <, ..., *<libref.catalog.>layer-entry-n*>) | _ALL_

adds the specified layer(s) to the LAYERSON list and deactivates any on-scale/off-scale settings for the specified layer(s).

LAYERSON+=(*<libref.catalog.>layer-entry-1* <, ..., *<libref.catalog.>layer-entry-n*>)
specifies a list of GISLAYER-type catalog entries that will be turned on for this map. All other layers will be turned off. Any on-scale/off-scale settings are deactivated. Specifying LAYERSON=(_ALL_) turns all layers on.

LAYERSON-=(*<libref.catalog.>layer-entry-1* <, ..., *<libref.catalog.>layer-entry-n*>)
removes the specified layer(s) from the LAYERSON list and deactivates any on-scale/off-scale settings for the specified layer(s).

LAYERSOFF=(*<libref.catalog.>layer-entry-1* <, ..., *<libref.catalog.>layer-entry-n*>) | _ALL_
specifies a layer (or list of layers) to be turned off for this map. All other layers are turned on. Any on-scale/off-scale settings are deactivated. Specifying LAYERSOFF=(_ALL_) turns all layers off.

LAYERSOFF+=(*<libref.catalog.>layer-entry-1* <, ..., *<libref.catalog.>layer-entry-n*>)
adds the specified layer(s) to the LAYERSOFF list and deactivate any on-scale/off-scale settings for the specified layer(s).

LAYERSOFF-=(*<libref.catalog.>layer-entry-1* <, ..., *<libref.catalog.>layer-entry-n*>)
removes the specified layer(s) from the LAYERSOFF list and deactivates any on-scale/off-scale settings for the specified layer(s).

Note: The following information applies to the LAYERSON and LAYERSOFF options:

- If a layer in any of the lists does not exist in the map, a warning is issued and that layer is ignored. (A missing layer does not end the current RUN-group processing.) Each layer is evaluated individually, so if other layers are valid they are toggled appropriately.
- If a layer is in both the LAYERSON list and the LAYERSOFF list, this condition generates a warning and ends that RUN-group.
- If one of the LAYERS options is specified in addition to LAYERSON or LAYERSOFF, the LAYERS parameters are processed first. Therefore, if a layer is removed from the map by using the LAYERS parameter, it cannot be referenced in a LAYERSON/LAYERSOFF parameter in that same statement. This action generates a warning, but the RUN-group processing does not stop.
- If both LAYERSON and LAYERSOFF are used in the same statement, both parameters must specify -=, +=, or both. Specifying both LAYERSON=(...) and LAYERSOFF=(...) in the same statement causes a conflict, and therefore is not allowed.
- The _ALL_ option cannot be mixed with layer names, that is, _ALL_ must appear by itself.
- _ALL_ cannot be used with either the += or the -= operators.

△

LEGEND=HIDEALL | UNHIDEALL | REMOVALL
used in conjunction with the MAP UPDATE statement to hide, display, or remove map legends:

HIDEALL
causes all existing legends to be hidden (not displayed) when the map is opened.

UNHIDEALL
causes all existing legends to be displayed when the map is opened.

REMOVEALL
removes all of the existing legends from the map.

CAUTION:

This behavior is immediate and permanent. You cannot restore the legends and will have to recreate them. △

Only one of the LEGEND= options can be specified at a time.

MULT=***multiplier-value***
specifies a constant integer value by which spatial data coordinates are multiplied when the data are displayed. The default is MULT=1E7. If the unit multiplier is too large, it is recomputed when the map is opened, and a note is printed to the SAS log showing the new value. If your map opens and appears to be empty, your MULT value might be too small.

NOWARN
specifies that messages are not to be issued about actions or attribute links that are not found during deletion. Use this argument when you specify the DELETE operation in the ACTION or ATTRIBUTE argument.

RENAME_LAYER ***old-name = new-name***
changes the name of an existing layer in the map that is being updated. This argument also changes the name of the layer entry in the catalog.

If other maps use the renamed layer, you must issue a MAP UPDATE statement for those maps as well.

SELECT=(*<libref.catalog.>layer-entry-1* <, ..., *<libref.catalog.>layer-entry-n*>>)
lists the layers to be selectable when the map opens. All other layers will be unselectable.

SELECT+=(*<libref.catalog.>layer-entry-1* <, ..., *<libref.catalog.>layer-entry-n*>)
adds layers to the current list of selectable layers.

SELECT-=(*<libref.catalog.>layer-entry-1* <, ..., *<libref.catalog.>layer-entry-n*>)
removes layers from the current list of selectable layers.

UNSELECT=(*<libref.catalog.>layer-entry-1* <, ..., *<libref.catalog.>layer-entry-n*>)
lists the layers to be unselectable when the map opens. All other layers will be selectable.

UNSELECT+=(*<libref.catalog.>layer-entry-1* <, ..., *<libref.catalog.>layer-entry-n*>)
adds layers to the current list of unselectable layers.

UNSELECT-=(*<libref.catalog.>layer-entry-1* <, ..., *<libref.catalog.>layer-entry-n*>)
removes layers from the current list of unselectable layers.

Details

A map entry is a SAS catalog entry of type GISMAP that defines the displayed features of a map. The definition specifies which layers the map contains and which coverage of the spatial database is used. The map entry also stores legend definitions and action definitions for the map, information about the projection system used to display the map, the name of the data set that contains labels for map features, and the names of any other associated SAS data sets.

In the MAP statement, the *map-entry* identifies the map entry you want to create, delete, replace, or update. The general form of the argument is the following:

<libref.catalog.>map-entry

If you specify a one-level name, the map entry is assumed to be in the catalog that is specified in the PROC GIS statement or in the most recently issued CATALOG statement. An error occurs if no catalog has previously been specified.

The *map-entry* name must conform to the rules for SAS names:

- □ The name can be no more than 32 characters long.
- □ The first character must be a letter or underscore (_). Subsequent characters can be letters, numeric digits, or underscores. Blanks are not permitted.
- □ Mixed-case names are honored for presentation purposes. However, because any comparison of names is not case-sensitive, you cannot have two names that differ only in case (for example, State and STATE are read as the same name).

Examples

Define a New Map

The following code fragment creates an entry named STORES of type GISMAP in the current catalog. The map is based on the coverage defined in the GISCOVER entry named MALL in the current catalog and uses the GISLAYER entries STORE, FIRE, INFO, PHONE, and RESTROOM in the current catalog.

```
map create stores / coverage=mall
                    layers=(store, fire, info, phone, restroom);
run;
```

Update an Existing Map Definition

The following code fragment updates the MAPS.USA.USA.GISMAP entry to use detail data when the map is displayed:

```
map update maps.usa.usa / details;
run;
```

Copy Attribute Data Set Links

The following code fragment copies the SIMPLUSR attribute link from GISSIO.SIMPLUS.SIMPLE to WORK.SIMPLE.SIMPLE:

```
proc gis;
   map update work.simple.simple /
       attribute = (name=simplusr
                    copy from=gissio.simplus.simple);
   run;
```

MAPLABEL Statement

Applies, modifies, or deletes labels on a map.

MAPLABEL<*operation*> <*option*>

Operations

CONTENTS

prints label information to the Output window. If you specify

- □ _ALL_, every label not associated with a layer is printed.

- TEXT='*string*', only labels matching that text are printed. The comparison is case sensitive.
- ROW=*integer*, only the label at that row is printed.

If no labels are output, a NOTE is printed to the log.

CREATE
creates a new label or labels. Unlike CREATE operations for other PROC GIS statements, duplicate labels are allowed.

DELETE
removes the specified labels and, depending on which optional arguments are specified, possibly deletes the label data set. The only valid optional arguments for DELETE are DATASET=, MAP=, LAYER=, TEXT=, IMAGE=, ROW=, and _ALL_. Any others will be ignored.

If you specify

- DATASET=*data-set-name* as the only argument, the label data set is deleted.
- MAP=*map-entry* as the only argument, the label data set reference is removed from the map entry, and the data set is deleted. If you do not specify MAP=, and all of the rows in the label data set are deleted, you are cautioned that any maps using the deleted data set will generate a WARNING when opened.
- ROW=, only the label at that data set row is deleted. ROW= and _ALL_ are not allowed together. If you use ROW= and TEXT=, the TEXT= is ignored and the label at that row is deleted.
- TEXT='*string*', every label literal having this exact string is removed from the label data set.

  Note: Literal labels are those not associated with a specific layer. △
- IMAGE=, the specified image is deleted.
- _ALL_, every label that is not associated with a layer is removed from the label data set.

Either DATASET= or MAP= is required or no deletions can occur.

Any DELETE operation that deletes all of the rows in the label data set also causes the data set to be deleted. If a data set is deleted, a NOTE is printed to the log. If MAP= is present and the label data set is deleted, the reference to the data set within the map entry is removed.

A note is printed in the log upon completion of a successful deletion.

REPLACE
replaces an existing label specified by TEXT='*string*' of ROW=. If the label does not exist, a CREATE is performed.

UPDATE
modifies an existing label.

Options

ALL
affects the behavior of the following operations:

- In a CONTENTS operation, _ALL_ prints every label that is not associated with a layer to the Output window.
- In a DELETE operation, _ALL_ deletes every label not associated with a layer.

ALL has no effect on CREATE, REPLACE, or UPDATE operations. If _ALL_ is detected, it is ignored.

ALL cannot be used in the same statement with ROW= or TEXT= options.

ATTACH_TO=MAP | WINDOW
controls the label positioning and is a required argument.

MAP
The label is attached to a location on the map. As you move the map in the window, the label moves with the map.

WINDOW
The label is attached to the map window. It remains fixed relative to the window as you move the map in the window.

COLOR=*color-name* | CX*rrggbb*
specifies the text color. The default color is BLACK.

color-name
is a SAS color-name, for example GREEN or RED.

CX*rrggbb*
is an RGB color, for example `CX23A76B`.

For more information about color-naming schemes, see "SAS/GRAPH Colors" in *SAS/GRAPH: Reference*

DATASET=<*libref.data-set-name*
specifies the label data set to which new labels are appended. If the data set does not exist, it is created.

If you specify a one-level data set name, the WORK library is assumed. If you specify both DATASET= and MAP=, and the map already references a label data set, the data set names are compared. If they are not the same and FORCE was not specified, a warning is printed, and the run group is terminated.

FONT=*fontname* | DEFAULT

fontname — specifies the font for the label text.

The following are some examples:

```
FONT = 'Times New Roman-12pt-Roman-Bold'
FONT = 'Display Manager font'
FONT = 'Sasfont (10x15) 10pt-9.7pt-Roman-Normal'
```

If the specified font is not found when the map is opened, the default system font is substituted and a note is printed to the log.

DEFAULT — assigns the default font to the label.

FORCE
replaces the existing label data set reference in a map when both DATASET= and MAP= are specified. If the map already references a label data set, its data set name is compared to the name specified with DATASET=. If they are not the same, the FORCE option causes the map's label data set reference to be overwritten and a note printed to the log. The map's original label data set is not deleted.

FRONT | BACK

FRONT — causes an image label to be drawn over the map features. This is the default for image labels.

BACK — causes an image label to be drawn beneath the map features.

These options do not apply to text labels.

IMAGE=*'pathname'* | *<libref.>catalog.entry<.image>* | *fileref*

specifies the location of an image to use as an image label on the map.

'pathname'

enables you to enter a host directory path to an image file. For example,

```
IMAGE='C:\My SAS Files\photo.gif
```

<libref.>catalog.entry<.image>

uses a **.IMAGE** type catalog entry for the image label. If you omit the library name from the statement, the WORK library is the default.

fileref

enables you to use an external file for an image label by specifying an active SAS fileref. The host directory path for this fileref is written to the label data set, not to the fileref.

MAP=*<libref.catalog>map-entry*

indicates the map entry on which to display the labels. If you indicate a one-level name, the map entry is assumed to be in the catalog that is specified in the PROC GIS statement or in the most recently issued CATALOG statement.

If the specified map entry already references a label data set, new labels are appended to that data set.

If the map entry does not reference a label data set, you must provide a label data set name with the DATASET= option. The labels are written to that data set, and the data set is then assigned to the specified map.

For the MAPLABEL statement, either MAP= or DATASET= is required. If both are present and the map already has a label data set, its name must match the name specified in the DATASET= option. One exception is if the FORCE option is also used. In that case, the label data set specified by DATASET= is assigned to the map entry. The map's original label data set is unassigned but not deleted.

OFFSET=(*<x>* <, *y*> <<,> *units*> | *<x* <*units*>> <, *y units*>>)

specifies the distance to move the entire label. **X** is the number of units to move the label right (positive numbers) or left (negative numbers), and **Y** is the number of units to move the label up (positive numbers) or down (negative numbers). For example:

To set only the **X** offset, specify one value, with or without a following comma:

```
OFFSET = ( 10 real, )
```

To set both the **X** and **Y** offset, specify two values, with or without a comma separating them:

```
OFFSET = ( 20 pct, 40 pct )
```

To set only the **Y** offset, specify one value preceded by a comma:

```
OFFSET = ( ,-30 pct )
```

OFFSET= is usually used in conjunction with POSITION= to adjust the position of a label. Moves are relative to the location specified by POSITION=, with OFFSET=(0,0) representing the initial position. You can also apply OFFSET= to the default label position.

The following restrictions apply to the OFFSET= argument:

- When both ORIGIN and OFFSET are used, the same units must be specified for both. If no units are specified in the OFFSET= arguments, the unit entered in the ORIGIN= option is used.
- OFFSET= is unnecessary with ORIGIN= since ORIGIN= explicitly positions the label and requires no further adjustment. However, if you specify both options,

the values of OFFSET= are added to the values of ORIGIN=, and the label is positioned accordingly.

- If the resulting location is outside of the GIS window, a WARNING is printed to the log when the map is opened, and the label is moved to be within the window.

ONSCALE=(scale <*units*>)

scale
specifies a map scale where the label is turned on or off when the map view is zoomed.

units
specifies the units for ONSCALE.

real-units/map-units
enables you to specify various combinations of units. Valid values are KM, M, CM, MI, FT, and IN. *Real-units* is typically KM, M, MI, or FT, and *map-units* is usually either CM or IN. Long forms of the unit names, for example KILOMETERS or INCH (singular or plural), are also acceptable.

METRIC
sets the scale units to KM/CM. This is the default if units are omitted.

ENGLISH
sets the scale units to MI/IN.

OFFSCALE=(scale <*units*>)

scale
specifies a map scale where the label is turned on or off when the map view is zoomed.

units
specifies the units for OFFSCALE.

real-units/map-units
enables you to specify various combinations of units. Valid values are KM, M, CM, MI, FT, and IN. *Real-units* is typically KM, M, MI, or FT, and *map-units* is usually either CM or IN. Long forms of the unit names, for example KILOMETERS or INCH (singular or plural), are also acceptable.

METRIC
sets the scale units to KM/CM. This is the default if units are omitted.

ENGLISH
sets the scale units to MI/IN.

ORIGIN=(<*x*> <, *y*> <,> <*units*> | <*x* <*units*>> <, *y units*>>)
specifies the `X` and `Y` coordinates for the label. ORIGIN= explicitly positions the label anywhere on the map. Unlike the POSITION= location, the label is not centered about this point. The lower left corner of the label is placed at the specified ORIGIN location.

- *x* indicates the X-coordinate (horizontal axis)
- *y* indicates the Y-coordinate (vertical axis)

Use the following units when attaching labels to the map:

REAL
`X` and `Y` values are real-world coordinates based on the underlying spatial data. Negative values can be used to signify the western or southern hemispheres if the spatial data contains them.

PERCENT | PCT
: refers to a percentage of the distance from the lower left corner of this bounding box to the label origin. All SAS/GIS maps have a rectangular bounding box that encompasses the extreme limits of the spatial data as subset by the GIS coverage. When a map is first opened in the GIS Map window, the map scale is set to display this entire box in the window. When the map is resized, the label remains in the same relative location, for example, for x=50, y=50 the label origin remains in the center of the map. Negative values and values greater than 100% can be used to place the origin outside of the coverage extents.

Use the following units when attaching labels to the window:

PIXEL
: **X** and **Y** values are screen coordinates using pixels. The lower left corner of the window is 0, 0. Negative values are not allowed. If a specified pixel value runs the label outside of the window, the label is shifted horizontally, vertically, or both to be just within the window when the map is opened.

PERCENT | PCT
: **X** and **Y** values are a percentage of the distance from the lower left corner of the map window to the label origin. When the window is resized, the label remains in the same relative location. For example, for x=50, y=50 the label origin remains in the center of the window. Negative values and values greater than 100% are not allowed.

To set only the **X** coordinate, specify one value with or without a following comma. For example:

```
ORIGIN = ( 10 pixels, )
```

or

```
ORIGIN = ( 10 pixels )
```

To set both the **X** and **Y** coordinates, specify two values with or without a comma separating them. The units can be specified for both **X** and **Y** or once at the end. For example:

```
ORIGIN = ( 10 pct, 40 pct )
```

or

```
ORIGIN = ( 10 pct  40 pct )
```

or

```
ORIGIN = ( 10  40  pct )
```

or

```
ORIGIN = ( 10, 40, pct )
```

To set only the **Y** coordinate, specify one value preceded by a comma. For example:

```
ORIGIN = ( , 20 pct )
```

ORIGIN= overrides the POSITION= option if both options are present. Although using the OFFSET= option with the ORIGIN= option is unnecessary, if you also specify OFFSET=, it is applied after the ORIGIN= request has been processed.

If the specified origin or origin plus offset is outside of the overall map bounding box, a warning is printed to the log. No warning is issued if the label runs out of the box, however.

OVERLAP | NOOVERLAP
specifies how labels are treated when they overlap.

OVERLAP
All labels you create with the option will be displayed even if they conflict with other labels.

NOOVERLAP
some of the conflicting labels will be suppressed until you zoom in more closely. This is the default setting.

POSITION = (<TOP | MIDDLE | BOTTOM | LEFT | CENTER | RIGHT>)
assigns a single label to a position on the map relative to the map's bounding box.

If the label is attached to the map, the label position is determined using the spatial data bounding box, that is the upper, lower, right and left edges of the coverage extents. If the label is attached to the window, the window edges are used.

The following arguments determine the vertical position of the label:

TOP
places the label along the upper edge of the map or window.

MIDDLE
places the label halfway between the top and bottom edges of the map or window.

BOTTOM
aligns the label with the bottom edge of the map or window.
The following arguments determine the horizontal position of the label:

LEFT
starts the label at the left edge of the map or window.

CENTER
places the label halfway between the left and right edges of the map or window.

RIGHT
aligns the end of the label with the right edge of the map or window. The label is centered horizontally and vertically at the computed position point. You can adjust the initial label position with the OFFSET= option.

ROW=*integer*
specifies a particular label in the data set to UPDATE, REPLACE, DELETE or print CONTENTS. Not valid for CREATE operations.

TEXT=<'*string*'
specifies the text for a literal label.

For REPLACE, UPDATE, DELETE or CONTENTS operations, *string* specifies a search string used to locate a specific target label. When TEXT='*string*' is used to search for a label, *string* is case sensitive, and an exact match to the value of the search string must be found. For example, TEXT='Paris' will ignore a label having PARIS as its text. It will also ignore a label having 'Paris Metro' as its text.

If ROW= and TEXT= are both present in a REPLACE or UPDATE statement the '*string*' entered with TEXT= is not a search string. It becomes the new text for the label at the specified ROW number.

TRANSPARENT | NOTRANSPARENT

TRANSPARENT
enables the map features to show though the label's bounding box. This is the default if not specified.

NOTRANSPARENT
blocks the display of map features within the label's rectangular bounding box.

These options apply to text labels and image labels stored as catalog entries. They have no effect on image labels stored in an external file.

COPY Statement

Copies a SAS/GIS catalog entry or data set. You can copy a single GIS entry or include the dependent entries and data sets that are referenced by the source.

COPY <*libref.catalog.*>*entry*<*.type*> </ *options*> ;

Options

ALIAS=(*old-libref-1=new-libref-1* <, ... ,*old-libref-n=new-libref-n*>)
specifies libref translations. The *old-libref* value is the libref that is stored in the existing catalog entry. The *new-libref* value is the libref that you want to substitute in the new copy of the entry.

BLANK
specifies that internal pathnames should be cleared in the copied entries.

DESTCAT=*libref.catalog*
specifies the destination for the copied catalog entries.

If *libref* is omitted, WORK is used as the default. Entries are copied to WORK.*catalog*. If DESTCAT= is omitted, *libref* defaults to WORK and the *catalog* to the catalog name of the source being copied. For example, if you are copying MAPS.USA.STATE, and you omit DESTCAT=, the copy of the data set is written to WORK.USA.STATE.

DESTLIB=*libref*
specifies the destination library for the copied data sets.

If DESTLIB= is omitted, the default *libref* is WORK.

ENTRYTYPE=*type*
specifies the type of GIS catalog entry to copy. The following are the values for *type*:

- GISSPA or SPATIAL
- GISMAP or MAP
- GISLAYER or LAYER
- GISCOVER or COVERAGE

This argument can be omitted if a complete, four-level entry name is specified. The following statement are identical:

- COPY MAPS.USA.STATE ENTRYTYPE=GISMAP
- COPY MAPS.USA.STATE.GISMAP

Note: When you specify four-level entry names, *type* must be the actual SAS/GIS catalog entry extension, for example, GISMAP, not MAP. △

REPLACE
specifies that both existing catalog entries and data sets that have the same name as copied entries and data sets should be overwritten.

SELECT= _ALL_ | ENTRY | DATASETS | LABEL | OTHER | NOSOURCE | SPATIAL

specifies which data sets or catalog entries that are referenced by the source entry should be copied. The following are the values for this option:

ALL
: copies all dependent catalog entries and data sets. It is equivalent to specifying both ENTRY and DATA.

ENTRY
: copies all dependent catalog entries.

DATASETS
: copies all dependent data sets. It is equivalent to specifying SPATIAL, LABEL, and OTHER.

LABEL
: copies dependent label data sets.

NOSOURCE
: copies entry dependents as specified, but does not copy the specified source entry.

OTHER
: copies other dependent data sets (besides spatial and label data sets), such as linked attribute data sets.

SPATIAL
: copies dependent spatial data sets.

Details

When you use the COPY statement, you must specify the catalog entry for the source entry you want to copy. If you specify a one-level name, the current catalog is used.

Note: When you specify a four-level entry name, the type must be the actual SAS/GIS catalog entry extension, for example, GISMAP instead of MAP. △

Note: If you use PROC COPY or another utility to copy a SAS/GIS catalog entry or data set, you might receive warnings in your SAS log that the paths are not the same. If you receive a message that the paths are not the same, you can use the SYNC statement to reset the paths. See "SYNC Statement" on page 150 for more information. △

MOVE Statement

Moves a SAS/GIS catalog entry or data set. You can move a single GIS entry or include the dependent entries and data sets that are referenced by the source.

Note: Before you use the MOVE statement on a catalog entry or data set, make sure that you have WRITE permission to the source location. The MOVE statement deletes the original entry or data set and creates a new copy in the target directory. If you do not have WRITE permission to the source location, MOVE leaves the original entry or data set in its directory and creates a copy in the target directory.

MOVE <*libref.catalog.*>*entry*<*.type*> </ *options*>;

Options

ALIAS=(*old-libref-1=new-libref-1* <, ... ,*old-libref-n=new-libref-n*>)
specifies libref translations. The *old-libref* value is the libref that is stored in the existing catalog entry. The *new-libref* value is the libref that you want to substitute in the moved entry.

BLANK
specifies that internal pathnames should be cleared in the moved entries.

CHECKPARENT
specifies that data sets and catalog entries are checked before they are moved to see what other GIS entries references them. If any references are found, the catalogs and data sets are copied instead of being moved.

If CHECKPARENT is not specified (the default), data sets and catalog entries are moved without checking for references, which might cause problems with other GIS entries.

CAUTION:
Do not use host commands to move or rename SAS data sets that are referenced in GISSPA entries. Moving or renaming a data set that is referred to in a spatial entry breaks the association between the spatial entry and the data set. To prevent breaking the association, use the PROC GIS MOVE statement with the CHECKPARENT option instead of a host command. △

DESTCAT=*libref.catalog*
specifies the destination for the moved catalog entries.

If *libref* is omitted, WORK is used as the default. Entries are moved to the WORK.*catalog*. If DESTCAT= is omitted, *libref* defaults to WORK and the *catalog* to the catalog name of the source being moved. For example, if you are moving MAPS.USA.STATE, and you omit DESTCAT=, the data set that you are moving is written to WORK.USA.STATE.

DESTLIB=*libref*
specifies the destination library for the moved data sets.

If DESTLIB= is omitted, the default libref is WORK.

ENTRYTYPE=*type*
specifies the type of GIS catalog entry to move. The following are the values for *type*:

- GISSPA or SPATIAL
- GISMAP or MAP
- GISLAYER or LAYER
- GISCOVER or COVERAGE

This argument can be omitted if a complete, four-level entry name is specified. The following statements are identical:

- `MOVE MAPS.USA.STATE ENTRYTYPE=MAP`
- `MOVE MAPS.USA.STATE.GISMAP`

Note: When you specify four-level entry names, *type* must be the actual SAS/GIS catalog entry extension, for example, GISMAP, not MAP. △

REPLACE
specifies that both existing catalog entries and data sets that have the same name as moved entries and data sets should be overwritten.

SELECT= _ALL_ | ENTRY | DATASETS | SPATIAL | LABEL | OTHER | NOSOURCE

specifies which data sets or catalog entries that are referenced by the source entry should be moved. The following are the values for this argument:

ALL
: moves all dependent catalog entries and data sets. Equivalent to specifying both ENTRY and DATA.

DATA
: moves all dependent data sets. It is equivalent to specifying SPATIAL, LABEL, and OTHER.

ENTRY
: moves all dependent catalog entries.

LABEL
: moves dependent label data sets.

NOSOURCE
: moves entry dependents as specified, but does not move the specified source entry.

OTHER
: moves other dependent data sets (besides spatial and label data sets), such as linked attribute data sets.

SPATIAL
: moves dependent spatial data sets.

Details

When you use the MOVE statement, you must specify the catalog entry for the source entry you want to copy. If you specify a one-level name, the current catalog is used.

Note: When you specify a four-level entry name, the type must be the actual SAS/GIS catalog entry extension, for example, GISMAP instead of MAP. △

Note: If you use PROC COPY or another utility to move a SAS/GIS catalog entry or data set, you might receive warnings in your SAS log that the paths are not the same. If you receive a message that the paths are not the same, you can use the SYNC statement to reset the paths. See "SYNC Statement" on page 150 for more information. △

SYNC Statement

Updates a SAS/GIS catalog entry libref and the internal pathname.

Note: You can sync a single GIS entry or include the dependent entries on a catalog entry. Before you use the SYNC statement on a catalog entry, make sure that you have WRITE permission to the source location.

SYNC <*libref.catalog.*>*entry*<*.type*> </ *options*>;

Options

ALIAS=(*old-libref-1=new-libref-1* <, ... ,*old-libref-n=new-libref-n*>)
specifies libref translations. The *old-libref* value is the libref that is stored in the existing catalog entry. The *new-libref* value is the libref that you want to substitute in the synchronized version of the entry.

BLANK
specifies that internal pathnames should be cleared in the updated entries.

ENTRYTYPE=*type*
specifies the type of GIS catalog entry to move. The following are the values for *type*:

- GISSPA or SPATIAL
- GISMAP or MAP
- GISLAYER or LAYER
- GISCOVER or COVERAGE.

This argument can be omitted if a complete, four-level entry name is specified. The following statements are identical:

- SYNC SASUSER.MALL.STORES ENTRYTYPE=GISMAP...
- SYNC SASUSER.MALL.STORES.GISMAP ...

Note: When specifying four-level entry names, *type* must be the actual SAS/GIS catalog entry extension, for example, GISMAP, not MAP. △

SELECT= _ALL_ | ENTRY
specifies which catalog entries that are referenced by the source entry should be updated. The following are the values for this argument:

ALL updates all dependent catalog entries. This is equivalent to specifying ENTRY.

ENTRY updates all dependent catalog entries.

PART 2

Appendixes

APPENDIX 1

Maps Supplied with SAS/GIS Software

Map and Data Sets Supplied with SAS/GIS Software

Several SAS/GIS sample maps and their associated data sets are supplied with SAS/GIS Software. These maps reside in the MAPS library, along with map data sets shipped with SAS/GRAPH Software. These maps can be used for exploring the software or for demonstration purposes.

To open the sample maps, invoke SAS/GIS and select **File ▸ Open Map** or right-click on the GIS Map window and select **Open Map**. Select the MAPS library, followed by the catalog and the map name.

Three sets of sample maps are included with SAS/GIS Software, and two maps are created by the SAS/GIS Tutorial. The following list provides the name of the map, and the spatial data sets, catalogs, and data sets associated with each map. The format of the map name is MAPS.*catalog.map-name*.

Additional maps can be created using the SAS/GIS import process.

Maps in the USA Catalog

These maps are all defined in the same spatial database; that is, they all use the same spatial data even though the maps represent different geographic areas.

MAPS.USA.STATE

This map includes USA and STATE boundaries for the continental United States, along with a CAP_CITY layer which represents the locations of state capitals. The theme defined for the STATE layer depicts average household income by state. A Browse action named BRSTATE displays information about the selected state. A Drill action defined for this map allows users to drill down to an individual county map of the selected state. Selecting the state of North Carolina and running the DRILL action opens a sample map of North Carolina by COUNTY and ZIP code (MAPS.NC.NC).

MAPS.USA.COUNTY

This map includes USA, STATE, and COUNTY boundaries for the continental United States. The theme defined for the COUNTY layer depicts population change by county. There are three actions defined for this map. A Browse action named BRCOUNTY displays information about the selected county. The TREND Program action creates a bar chart that displays population and housing trends. A Program action named RATIO creates a plot that displays population and housing ratios for the selected county.

MAPS.USA.TRAINING

This map includes USA and STATE boundaries for the continental United States, along with a CENTER point layer which represents the locations of SAS Training Centers. This map contains an IMAGE action which displays a picture of the selected training center. There is also a Browse action named BRCENTER which displays information about the selected training center.

MAPS.USA.*state*

(where *state* is the two-character postal code for the state)

There is an individual map for every state in the continental United States, plus the District of Columbia. Each map is named using the two-character postal code for the particular state. These maps include USA, STATE, and COUNTY boundaries. The theme defined for the COUNTY layer is the same as in the MAPS.USA.COUNTY map, and depicts population change by county. There are three actions defined for the maps. A Browse action named COUNTY displays information about a selected county. The TREND Program action creates a bar chart that displays population and housing trends. A Program action named RATIO creates a plot that displays population and housing ratios for a selected county.

Associated catalog and data sets:

USA	Catalog
USAC	Chains data set
USAN	Nodes data set
USAD	Details data set
USACTI	Polygonal index data set for the COUNTY layer
USASTI	Polygonal index data set for the STATE layer
USACTLAB	Label data set for the COUNTY map
USALAB	Label data set for the TRAINING map
USASTLAB	Label data set for the STATE map
USAAC	Attribute data set for the COUNTY map
USAAS	Attribute data set for the STATE map
USAAT	Attribute data set for the TRAINING map

Maps in the NC Catalog

MAPS.NC.NC

This map consists of COUNTY and ZIP code boundaries in North Carolina, along with a CITY layer representing the locations of major cities. A Drill action defined for this map enables users to drill down to a map of Wake County (MAPS.WAKE.TRACT) by selecting Wake County and running the DRILL action. This map contains three additional actions that can be run for the selected ZIP code:

- A Browse action that displays population and household data.
- A Program action named SUMMARY that executes a MEANS procedure and produces a summary for the ZIP code.
- A Program action named TREND that produces a line graph depicting population and housing trends.

Associated catalog and data sets:

NC	Catalog
NCC	Chains data set
NCN	Nodes data set
NCD	Details data set
NCCTI	Polygonal index data set for the COUNTY layer
NCZIPI	Polygonal index data set for the ZIP layer
NCCTLAB	Label data set for the CITY layer
NCLAB	Label data set for the CITY layer
NCWKLAB	Label data set for the COUNTY layer (Wake County)
NCAZ	Attribute data set

Maps in the WAKE Catalog

These maps are all defined in the same spatial database; that is, they all use the same spatial data even though the maps are not identical.

MAPS.WAKE.BG

This map includes COUNTY, TRACTS, and BG (block group) boundaries, as well as a STREET layer, for Wake County, North Carolina. The theme defined for the BG layer depicts the number of households by block group. This map includes a Browse action named CENSUS which displays information about the selected block group.

MAPS.WAKE.GROCERY

This map includes COUNTY and TRACT boundaries for Wake County, North Carolina, as well as a STREET layer and a GROCERY layer, which represents the locations of grocery stores. The theme defined for the TRACT layer depicts the average household income by tract. This map includes three actions: a Program action named GRAPH that produces a bar chart of store size versus sales for the selected store; a Browse action that displays information about the selected store; and a Spatial action that displays spatial information about the selected store.

MAPS.WAKE.TRACT

This map includes COUNTY and TRACT boundaries for Wake County, North Carolina, as well as a STREET layer. The theme defined for the TRACT layer depicts the neighborhood type by tract. This map includes a Browse action that displays demographic information about the selected tracts.

Associated catalog and data sets:

WAKE	Catalog
WAKEC	Chains data set
WAKEN	Nodes data set
WAKED	Details data set
WAKEBGI	Polygonal index data set for the BG layer
WAKETRTI	Polygonal index data set for the TRACTS layer
WAKELAB	Label data set
WAKEABG	Attribute data set for the BG map
WAKEAG	Attribute data set for the GROCERY map
WAKEAT	Attribute data set for the TRACT map

Copying and Modifying SAS/GIS Maps in the MAPS Library

Since most SAS Software users do not have write permission to the MAPS library, you will not be able to save any changes that you make to the SAS/GIS sample maps. You will see an error similar to the following in your SAS Log if you try to save any changes made to these maps:

```
ERROR: Write access to member MAPS.NC.CATALOG is denied.
```

In order to save modifications to these maps, you must first copy the map and its associated data sets to a SAS library to which you have write access. Use the Copy utility included with SAS/GIS to copy the maps.

For example, to copy the MAPS.NC.NC map and its associated data sets from the MAPS library to the SASUSER library, invoke SAS/GIS, select **Edit ▸ Copy**

Follow these steps:

1 Enter or browse and select MAPS.NC.NC for the Entry field.

2 Select Map as the entry Type.

3 In the Catalog Entry Destination field, enter the library and catalog to which you will be copying the map. You must specify a currently allocated SAS library. If the

specified catalog does not already exist in the library, it will be created during the copy process. In this example, enter SASUSER.NC.

4 In the Options box, select all options:
   - □ Copy Source Entry
   - □ Copy Dependent Entries
   - □ Copy Dependent Data Sets
   - □ Spatial Label Other
   - □ Replace Like-Named Entries/Data Sets
   - □ Blank out Path Name References

5 In the Data Set Destination field, enter the library to which the data sets associated with the map will be copied. In this case, enter SASUSER.

6 Click the Apply button to begin copying the map. When the copy is complete, you will see a note in the message line similar to the following:

   Number of entries, data sets copied: 9, 8

You will also see a list inthe SAS Log of all of the catalogs and data sets that were copied.

Now open the SASUSER.NC.NC map in the GIS Map window. You will now be able to make any desired modifications to the map, and save those modifications to the copied map.

Maps Produced by the SAS/GIS Tutorial

Two sample maps are produced by the SAS/GIS Tutorial. The maps are created automatically when the tutorial is invoked by selecting **Help ▸ Getting Started with SAS/GIS Software ▸ Begin Tutorial** from the GIS Map window. You can also create the maps without invoking the Tutorial by selecting **Help ▸ Getting Started with SAS/GIS Software ▸ Create Data**.

These maps are created in the SASUSER library, so changes made to these maps can be saved. The following list provides the name of the map, and the spatial data sets, catalogs, and data sets associated with each map. These two maps are defined in the same spatial database; that is, they use the same spatial data even though the maps are not identical.

SASUSER.MALL.AREA

This map includes TRACT boundaries for several tracts in Wake County, North Carolina. The map also includes point layers MALL, PARK, and SCHOOL. The map does not have a theme defined for the TRACT layer. A STREET layer is also included in this map. A Drill action defined for the MALL layer allows users to drill down to a detailed floor plan of a mall (SASUSER.MALL.STORES).

Associated catalog and data sets:

MALL	Catalog
MALLC	Chains data set
MALLN	Nodes data set
MALLD	Details data set

MALLTI	Polygonal index data set for the TRACT layer
MALLPOP	Attribute data set

SASUSER.MALL.STORES

This map consists of STORE and ATRIUM boundaries for a fictitious shopping mall. The map includes FIRE and SPRINK line layers, which depict fire exits and sprinkler systems, as well as the following point layers: INFO, PHONE, RESTROOM, STROLLER, SECURITY, and ALARM.

There are two actions defined for this map. The STOREPIC image action displays a picture of a store when one of the areas of the STORE layer is selected. The CHART Program action creates a bar chart of the square footage of the selected store(s).

Associated catalog and data sets:

MALL	Catalog
MALLC	Chains data set
MALLN	Nodes data set
MALLD	Details data set
MALLSI	Polygonal index data set for the STORE layer
MALLAI	Polygonal index data set for the ATRIUM layer
MALLSTOR	Attribute data set

APPENDIX 2

Details of SAS/GIS Spatial Databases

The SAS/GIS Data Model

SAS/GIS software uses two basic types of data:

spatial data
: describes the location, shape, and interrelationships of map features.

attribute data
: provides information that relates to the map features.

Spatial Data Features

SAS/GIS software uses spatial data to represent the following three types of map features:

point features
: consist of individual locations that are shown as symbols, representing real-world locations of special points of interest.

line features
: consist of sequences of two or more coordinates that form zero-width shapes, either closed or unclosed. Line features represent entities that either have no width, such as political boundaries, or those that can be represented as having no width, such as streets or water pipes.

area features
: consist of sequences of three or more coordinates that form polygons (with single or multiple boundaries and with or without holes.) Area features represent two-dimensional entities such as geographic areas (countries, states, and so on) or floor plans for buildings.

SAS/GIS Topology

To represent point, line, and area features in a map, SAS/GIS software defines the following topological features in the spatial data:

chains
: are sequences of two or more points in the coordinate space. The end points (that is, the first and last points of the chain) are nodes. Each chain has a direction, from the first point toward the last point. The first point in the chain is the *from-node* and the last point is the *to-node*. Relative to its direction, each chain has a left side and a right side.

  Points between the from-node and the to-node are *detail points*, which serve to trace the curvature of the feature that is represented by the chain. Detail points are not nodes.

nodes
: are points in the spatial data coordinate space that have connections to one or more chains.

areas
: are two-dimensional finite regions of the coordinate space. One or more chains, called *boundary chains*, separate two different areas. Chains that lie completely inside an area are called *internal chains* and are bounded on the left and right sides by the same area.

The spatial data coordinate space can be represented in any numeric units, even those that include arbitrary values. Coordinates that are stored as longitude and latitude values have a maximum usable precision of about one centimeter.

Representations of map features are implemented with one or more chains, as follows:

point features
: are implemented with one chain, one node (that is, the from-node and to-node for a point feature are the same node), and no detail points.

line and area features
: are implemented with one or more chains and one or more nodes.

Rules for Topological Correctness

SAS/GIS spatial data must obey the following rules in order for the topology to be correct. These rules are similar to the rules for TIGER files from the U.S. Census Bureau. For more information on these rules, see Gerard Boudriault's 1987 article, "Topology in the TIGER File" in *AUTO-CARTA 8, Proceedings*, pages 258-263, published by the American Society for Photogrammetry and Remote Sensing and the American Congress on Surveying and Mapping.

Topological Completeness

All chains must have the following characteristics:

- They must be bounded by two nodes, the from-node and the to-node.

  Note: In chains for point features, for single-chain closed-loop line features, or for area boundaries, the from-node and the to-node are the same node, but both are still included in the chain definition. △

- They must be bounded by two areas, one on the left and one on the right.

These relationships must be complete, so the following two rules apply:

- The sides of all the chains incident to any given node must form a cycle. A *cycle* consists of one or more chains that start or end at the same node.
- The end points of chains that bound an area must form one or more disjoint cycles.

For each unique area ID or unique set of area IDs, all the boundary chains that have the ID value (either on the right or left, but not both) form one or more closed loops or cycles.

Topological-geometric Consistency

The collection of chains, nodes, and areas must have coordinates that make the collection a disjoint partitioning of the coordinate space. The following four conditions must be true to avoid problems with displaying the spatial data:

- No two points in the combined set of nodes and detail points can share the same coordinate.
- No two line segment interiors can share a common coordinate.
- No two areas can share a common coordinate.

  Note: Graphically overlaid data can have overlapping polygons, chains, and nodes and have no topological interconnectivity △

- Polygons that form the boundaries of holes inside areas must fall completely within the enclosing areas.

Note: Edge-matched data shares coordinates along common boundaries, but each chain should have the proper polygonal ID values on the side that represents the outside edge of their respective physical coverages as well as on the inside edge. △

Problems Resulting from Topological Errors

Topological errors in the spatial data cause the following types of problems:

- A polygonal index cannot be built for all the polygons for a particular area set.
- A successfully indexed polygon does not close because of the following problems:
  - The chains for a node do not form a cycle, which is sometimes the result of left- and right-side values being swapped for one or more of the connected chains.
  - A chain crosses another chain's interior coordinated space.
- Multiple features are selected when only one selection is desired because of overlapping features in a coordinate space.
- Select → Like Connected processing fails to select apparently connected chains.

Attribute Data Features

Attribute data is all other data that is related to map features in some way, including the data that you want to analyze in the context of the map. Attribute data can be stored in the spatial database by the following methods:

- directly with the spatial data as variables in the chains data set
- indirectly in SAS data sets that are joined to the chains data set by a link that includes one or more variables.

Attribute data can be used as follows:

- as themes for map layers.
- by *actions* that display or manipulate the attribute data when features are selected in the map. Actions can be defined to display the attribute data, create new SAS data sets that contain subsets of the attribute data, or submit SAS programs to process the attribute data.

SAS/GIS Spatial Database Structure

A SAS/GIS spatial database consists of a set of SAS data sets that store the spatial data and a set of SAS catalog entries that define the functions of, and the relationships between, the spatial data elements.

Spatial Data Sets

As a component of SAS, SAS/GIS software stores all of its spatial data in SAS data sets. The data sets for a SAS/GIS spatial database work together as one logical file, even though they are split into multiple physical files.

The spatial data sets implement a network data structure with links that connect chains to their two end nodes and each node to one or more chains. This structure is implemented by using direct pointers between the nodes and chains data sets. The details data set provides curvature points between nodes of chains, while the polygonal index data set provides an efficient method of determining the correct sequence of chains to represent polygons.

Common Spatial Data Set Variables

The following spatial data variables appear in the chains, nodes, and details data sets:

Variable	Description
ROW	row number (used as a link when the spatial data set is used as a keyed data set as well as for database protection)
DATE	SAS datetime value when the record was last modified
VERSION	data version number
ATOM	edit operation number
HISTORY	undo history record pointer

Variable Linkages in the Spatial Data

The following linkages exist between and within the spatial data sets:

Data Set	Variable	Links to...
chains	ROW[1]	self
	FRNODE	this chain's from-node record in the nodes data set
	TONODE	this chain's to-node record in the nodes data set
	D_ROW	first detail record in the details data set for this chain
nodes	ROW	self
	C_ROW1-C_ROW5	records in the chains data set of chains that are using this node
	NC[2]	node record in the nodes data set used to store additional chain records
details	ROW	self
	C_ROW	parent chain record in the chains data set or next detail continuation record in the details data set
index[3]	C_ROW	record in the chains data set for the first chain in this polygon

1 The ROW variable is used as a link between records in the spatial data sets. The ROW variable value for the first record of a feature in the chains or nodes data sets is considered the *feature ID*. Because some records in the nodes data set are continuations of other records, not every row number in the nodes data set is a feature ID. As a result, node feature ID numbers are not necessarily sequential.The ROW variable also provides protection against corruption of the database that is caused by the accidental insertion or deletion of records. If records were linked by physical record number rather than by ROW value, an improper record insertion or deletion would throw off all linkages to subsequent records in the database. In the event the database is corrupted, the ROW variable can be used to move the records back into their proper locations with minimal data loss.

2 A negative value indicates that the variable points to a continuation record. The absolute value of the variable is the row number of the next record used for that feature's data. In newly imported spatial data, continuation records always point to the next record in the data set, but this is not required. New chains can be attached to existing nodes without having to insert records, which would require extensive printer reassignments.

3 The index data set has no ROW variable because it can be easily rebuilt from the chains, nodes, and details data sets from which it was originally constructed.

Because the data sets are linked together by row number, the chains, nodes, and details data sets must be radix-addressable and might not be compressed.

Chains Data Set

The chains data set contains coordinates for the polylines that are used to form line and polygon features. A *polyline* consists of a series of connected line segments that are chains. The chains data set also contains the information that is necessary to implement nodes in the database.

The following system variables are unique to the chains data set:

Variable	Description
FRNODE[1]	starting from-node record for the chain
TONODE	ending to-node record for the chain

Variable	Description
D_ROW	first detail point record
ND	number of detail points in the chain
RANK	sorting key used to sort all the chains around an arbitrary node by their angle, starting from 0 and proceeding counter-clockwise. See Appendix 3, "Calculating Chain Rank," on page 173 for information about sorting a chain around its to- and from-node and for examples of calculating the to-node value, from-node value, and chain rank.
XMIN	minimum X coordinate of chain
XMAX	maximum X coordinate of chain
YMIN	minimum Y coordinate of chain
YMAX	maximum Y coordinate of chain

1 The TONODE and FRNODE variables can point to the same record.

The XMIN, YMIN, XMAX, and YMAX variables define a bounding box for the chain. These variables are included in the chains data set to make it possible to select all the chains in a given X-Y region by looking only at the chains data set.

In addition to the system variables, the chains data set might contain any number of attribute variables, some of which might be polygon IDs. Because the chains have left and right sides, there are typically paired variables for bilateral data such as polygon areas or address values. The names of the paired variables typically end with **L** or **R** for the left and right sides, respectively. For example, the data set might contain COUNTYL and COUNTYR variables with the codes for the county areas on the left and right sides of the chain, respectively. However, this naming convention is not required.

Nodes Data Set

The nodes data set contains the coordinates of the beginning and ending nodes for the chains in the chains data set and the linkage information that is necessary to attach chains to the correct nodes. A node definition can span multiple records in the nodes data set, so only the starting record number for a node is a node feature ID.

The following system variables are unique to the nodes data set:

Variable	Description
C_ROW1- C_ROW5	chain records for the first five chains connected to the node. If fewer than five chains are connected to the node, the unused variables are set to 0.
NC	number of chain pointers (if five or fewer chains are connected to the node) or the negative of the next continuation node record number (if more than five chains are connected to the node). See "Variable Linkages in the Spatial Data" on page 165 for more information about how NC is used to string continuation node records.
X	X coordinate of node.
Y	Y coordinate of node.

Details Data Set

The details data set stores curvature points of a chain between the two end nodes. Therefore, it contains all the coordinates between the intersection points of the chains. The node coordinates are not duplicated in the details data set.

The following system variables are unique to the details data set:

Variable	Description
C_ROW	parent chain record (if the chain has ten or fewer detail points) or the negative of the next continuation detail record (if the chain has more than ten detail points). See "Variable Linkages in the Spatial Data" on page 165 for a description of how C_ROW is used to string continuation detail records.
X1-X10	X coordinates of up to 10 detail points.
Y1-Y10	Y coordinates of up to 10 detail points.

Detail coordinate pairs (X2, Y2) through (X10, Y10) contain missing values if they are not used. The missing values ensure that the unused coordinate pairs are never used in any coordinate range calculation. The various importing methods set unused detail coordinates to missing as a precautionary measure.

Polygonal Index Data Set

Polygonal indexes are indexes to chains data sets. The index contains a record for each boundary of each polygon that was successfully closed in the index creation process. The same rules that are used to construct polygons are also used to construct polygonal indexes.

The following system variables are unique to polygonal index data sets:

Variable	Description
C_ROW	starting chain from which a polygon can be dynamically traversed and closed. This chain is sometimes referred to as the *seed chain* for the polygon. Any chain on a polygon's boundary can be the seed chain.
FLAGS	control flag for polygons.
NC	number of chains in the polygon boundary.

Polygonal index data sets are created with the POLYGONAL INDEX statement in the GIS procedure. See "POLYGONAL INDEX Statement" on page 99 for more information about using the GIS procedure to create polygonal index data sets.

Catalog Entries

SAS/GIS software uses SAS catalog entries to store *metadata* for the spatial database—that is, information about the spatial data values in the spatial data sets.

Note: Using host commands to move, rename, or delete SAS/GIS catalogs entries can break internal linkages. For details of how to manage catalog entries, see the COPY, MOVE, and SYNC statements in Chapter 7, "The GIS Procedure," on page 85. △

SAS/GIS spatial databases use the following entry types.

Spatial Entries

A *spatial entry* is a SAS catalog entry of type GISSPA that identifies the spatial data sets for a given spatial database and defines relationships between the variables in those data sets.

Spatial entries are created and modified using the SPATIAL statement in the GIS procedure.

Note: You can also create a new spatial entry by making the following selections from the GIS Map window's menu bar:**File ▶ Save As ▶ Spatial** △

SAS/GIS software supports simple spatial entries and merged spatial entries as follows.

Simple spatial entries contain the following elements:

- references to the chains, nodes, and details data sets that contain spatial information.
- references to any polygonal index data sets that define the boundaries of area features in the spatial data.
- definitions for composite associations that specify how the variables in the spatial data sets are used. See "Composites" on page 171 for more information.
- the definition for a lattice hierarchy that specifies which area features in the spatial data enclose or are enclosed by other features.
- the parameters for the projection system that is used to interpret the spatial information that is stored in the spatial data sets.
- the accumulated bounding extents of the spatial data coordinates of its underlying child spatial data sets, consisting of the minimum and maximum X and Y coordinate values and the ranges of X and Y values.

Merged spatial entries have the following characteristics:

- consist of multiple SAS/GIS spatial databases that are linked together hierarchically in a tree structure.
- contain logical references to two or more child spatial entries. A child spatial entry is a dependent spatial entry beneath the merged spatial entry in the hierarchy.
- contain specifications of how the entries were merged (by overlapping or edgematching).
- do not have their own spatial data sets.
- reference the spatial data sets that belong to the child spatial entries beneath them on the hierarchy.
- do not have references to any polygonal index data sets that define the boundaries of area features in the spatial data.
- do not have definitions for composites that specify how the variables in the spatial data sets are used. See "Composites" on page 171 for more information about composites.
- do not have the definition for a lattice hierarchy that specifies which area features in the spatial data enclose or are enclosed by other features.
- do not have parameters for the projection system that is used to interpret the spatial information that is stored in the spatial data sets.

- contain the accumulated bounding extents of the spatial data coordinates of their underlying child spatial entries, consisting of the minimum and maximum X and Y coordinate values and the ranges of X and Y values.

Merged spatial entries can help you to manage your spatial data requirements. For example, you might have two spatial databases that contain the county boundaries of adjoining states. You can build a merged spatial entry that references both states and then you can view a single map containing both states' counties. Otherwise, you would have to import a new map that contains the two states' counties. This new map would double your spatial data storage requirements.

The following additional statements in the GIS procedure update the information in the spatial entry:

COMPOSITE statement
: creates or modifies composites that define the relation and function of variables in the spatial data sets. See "COMPOSITE Statement" on page 94 for details about using the GIS procedure to create or modify composites.

POLYGONAL INDEX statement
: updates the list of available index names that are stored in the spatial entry. See"POLYGONAL INDEX Statement" on page 99 for details about using the GIS procedure to create or modify polygonal indexes.

LATTICE statement
: updates the lattice hierarchy that is stored in the spatial entry. See "LATTICE Statement" on page 102 for details about using the GIS procedure to define lattice hierarchies.

You can view a formatted report of the contents of a spatial entry by submitting a SPATIAL CONTENTS statement in the GIS procedure.

See "SPATIAL Statement" on page 90 for more information about using the GIS procedure to create, modify, or view the contents of spatial entries.

Coverage Entries

A *coverage entry* is a SAS catalog entry of type GISCOVER that defines the subset, or *coverage*, of the spatial data that is available to a map. SAS/GIS maps refer to coverages rather than directly to the spatial data.

a coverage entry contains the following elements:

- a reference to the root spatial entry.
- a WHERE expression that describes the logical subset of the spatial data that is available for display in a map. (The expression WHERE=1 can be used to define a coverage that includes all the data that is in the spatial database.)

  The WHERE expression binds the coverage entry to the spatial data sets that it subsets. The WHERE expression is checked for compatibility with the spatial data when the coverage entry is created and also whenever a map that uses the coverage entry is opened.
- the maximum and minimum X and Y coordinates in the portion of the spatial data that meets the WHERE expression criteria for the coverage.

  These maximum and minimum coordinates are evaluated when the coverage is created. The GIS procedure's COVERAGE CREATE statement reads the matching chains and determines the extents from the chains' XMIN, YMIN, XMAX, and YMAX variables. If you make changes to the chains, nodes, and details data sets that affect the coverage extents, you should use the COVERAGE UPDATE statement to update the bounding extent values.

Multiple coverage entries can refer to the same spatial entry to create different subsets of the spatial data for different maps. For example, you could define a series of coverages to subset a county into multiple sales regions according to the block groups that are contained in each of the regions. The spatial data for the county would still be in a single spatial database that is represented by the chains, nodes, and details data sets and by the controlling spatial entry.

Coverage entries are created and modified using the COVERAGE statement in the GIS procedure. You can view a formatted report of the contents of a coverage entry by submitting a COVERAGE CONTENTS statement in the GIS procedure. (The contents report for a coverage entry also includes all the contents information for the root spatial entry as well.)

See “COVERAGE Statement” on page 105 for more information about using the GIS procedure to create, modify, or view the contents of coverage entries.

Layer Entries

A *layer entry* is a SAS catalog entry of type GISLAYER that defines the set of features that compose a layer in the map. A layer entry contains the following elements:

- a WHERE expression that describes the common characteristic of features in the layer.

  The WHERE expression binds the layer entry to the spatial data even though it is stored in a separate entry. The layer is not bound to a specific spatial entry, just to those entries representing the same type of data. Therefore, a layer that is created for use with data that is imported from a TIGER file can be used with data that is imported from any TIGER file; however, not all file types can take advantage of this behavior. The WHERE expression is checked for compatibility with spatial data when the layer entry is created and also whenever a map that uses the layer entry is opened.

  Note: When you define area layers, you can specify a composite as an alternative to specifying an explicit WHERE expression. However, the layer entry stores the WHERE expression that is implied by the composite. For example, if you specify STATE as the defining composite for a layer, and the STATE composite specifies the variable association VAR=(LEFT=STATEL,RIGHT=STATER), then the implied WHERE expression that is stored in the layer entry is WHERE STATEL NE STATER. △

- option settings for the layer such as the layer type (point, line, or area), whether the layer is static or thematic, whether it is initially displayed or hidden, whether detail points are drawn for the layer, and the scales at which the layer is automatically turned on or off.
- the graphical attributes that are necessary to draw the layer if it is a static layer.
- the attribute links, theme range breaks, and graphical attributes if the layer contains themes.

See “LAYER Statement” on page 107 for more information about using the GIS procedure to create, modify, or view the contents of layer entries.

Map Entries

A *map entry* is a SAS catalog entry of type GISMAP. Map entries are the controlling entries for SAS/GIS maps because they tie together all the information that is needed to display a map. A map entry contains the following elements:

- a reference to the coverage entry that defines the subset of the spatial data that is available to the map. Note that the map entry refers to a particular coverage of the spatial data rather than directly to the spatial entry.

- references to the layer entries for all layers that are included in the map.
- references to any attribute data sets that are associated with the map, along with definitions of how the attribute data sets are linked to the spatial data.
- a reference to the SAS data set that contains labels for map features.
- definitions for the actions that can be performed when map features are selected.
- definitions for map legends.
- parameters for the projection system that is used to project spatial data coordinates for display.
- option settings for the map, including the following:
  - the units and mode for the map scale
  - whether coordinate, distance, and attribute feedback are displayed
  - whether detail points are read
  - whether the tool palette is active.

Map entries are created using the MAP CREATE statement in the GIS procedure. However, much of the information that is stored in the map entry is specified interactively in the GIS Map window.

You can view a formatted report of the contents of a map entry by submitting a MAP CONTENTS statement in the GIS procedure. (The contents report for a map entry includes all the contents information for the spatial, coverage, and layer entries as well.)

See “MAP Statement” on page 129 for details about the items that can be specified with the GIS procedure. See Chapter 10, “SAS/GIS Windows” in *SAS/GIS Software: Usage and Reference, Version 6* for details about the items that can be specified interactively in the GIS Map window.

Composites

For most operations that involve the spatial database, you refer to composites of the spatial data variables rather than directly to the variables in the spatial data sets. A composite consists of the following elements:

- a *variable association* that identifies which variable or variables in the spatial database comprise the association. The variable association can specify a single variable, a pair of variables that define a bilateral (left-right) association, or two pairs of variables that define the start and end of a directional (from-to) bilateral association.
- a *class attribute* that identifies the role of the composite in the spatial database.

For example, if the chains data set has a variable that is named FEANAME that contains feature names, then you can create a composite for the FEANAME variable that assigns the class attribute NAME to indicate that the composite represents feature names. If the chains data set has COUNTYL and COUNTYR variables that contain the codes for the counties on the left and right sides of the chains, then you can create a composite that is named COUNTY. The composite identifies the bilateral relationship between these two variables and assigns the class attribute AREA to indicate that the composite defines county areas in the spatial data.

Composites are created and modified using the COMPOSITE statement in the GIS procedure. Composite definitions are stored in the spatial entry.

See “COMPOSITE Statement” on page 94 for more information about using the GIS procedure to create or modify composites.

APPENDIX

3

Calculating Chain Rank

RANK Value Equation

RANK is the sorting key used to sort multiple chains that have a common node by their angle, starting from 0 at due east and proceeding counterclockwise. A node can be either of the two end-points of a chain.

RANK values have the form *ffffff.tttttt*, where the *ffffff* value is used to sort the chain around its from-node and the *tttttt* component is used to sort the chain around its to-node. The *ffffff* and *tttttt* components are calculated using the following formula:

$$R = 1E5\left\{(Q - 1) + \tan\left[\frac{A}{2}\right]\right\}$$

R   is the calculated ranking factor.

Q   is the quadrant number (1 to 4, Figure A3.1 on page 174) that contains the angle α for the chain. For the *ffffff* component, α is defined by the vector $F \rightarrow D_0$, where F is the from-node and D_0 is the first detail point. For chains that have no detail points, D_0 is the to-node. For the *tttttt* component, α is defined by the vector $T \rightarrow D_L$, where T is the to-node and D_L is the last detail point. For chains that have no detail points, D_L is the from-node.

A is the angle from the chain clockwise to the nearest X or Y axis, and is determined with

$$\alpha - (Q - 1)\frac{\pi}{2}$$

where α is the clockwise angle from the chain to the positive x-axis (due east).

The tangent term is called the half-angle tangent. Since the angle A/2 can never exceed $\pi/4$ (45 degrees), the half-angle tangent has values from 0 to 1. The (Q-1) multiplier adjusts the range of values to 0 to 4. The values 0, 1, 2, 3, and 4 represent angles of 0, 90, 180, 270, and just under 360 degrees, respectively.

The 1E5 multiplier is used to transform decimal rank values to integers. Thus the rank values for a chain have six significant digits.

Note: The trigonometric functions are in radians. △

Calculating the Value of A in a Quadrant

The following figure illustrates the relationship of the quadrants to each other. Note that their numerical order is counterclockwise.

Figure A3.1 Quadrant Numbers

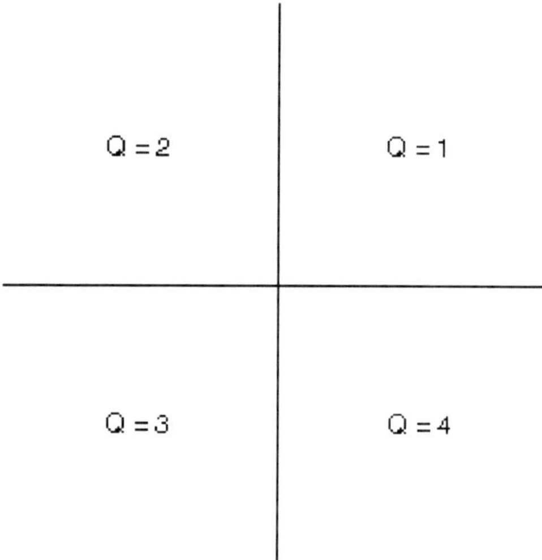

The following figures illustrate how to calculate the value of A in each quadrant.

The following calculations were used to determine the rank in Figure A3.2 on page 175:

given

$$R_F = 1E5\left[(Q - 1) + \tan\left(\frac{A}{2}\right)\right]$$

since

$$Q = 1$$

then

$$R_F = 1E5\left[\tan\left(\frac{A}{2}\right)\right]$$

Figure A3.2 Calculating Rank in Quadrant 1

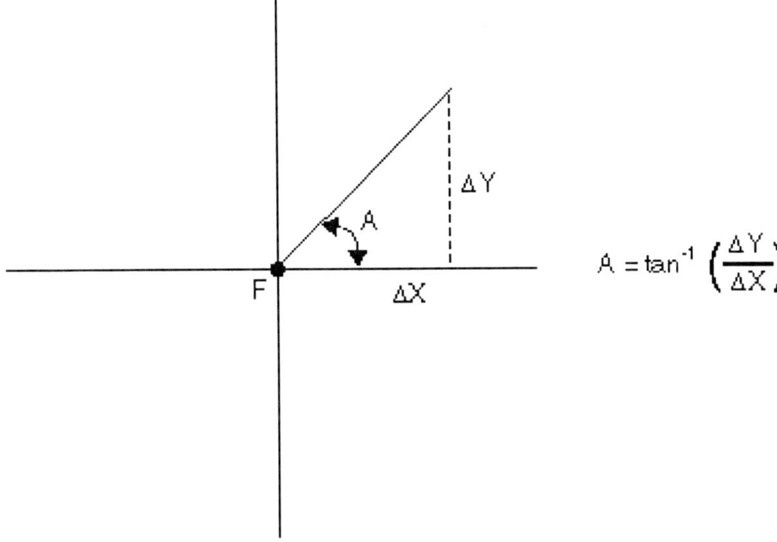

$$A = \tan^{-1}\left(\frac{\Delta Y}{\Delta X}\right)$$

The following calculations were used to determine the rank in Figure A3.3 on page 176:
given

$$R_F = 1E5\left[(Q-1) + \tan\left(\frac{A}{2}\right)\right]$$

since

$$Q = 2$$

then

$$R_F = 1E5\left[1 + \tan\left(\frac{A}{2}\right)\right]$$

Figure A3.3 Calculating Rank in Quadrant 2

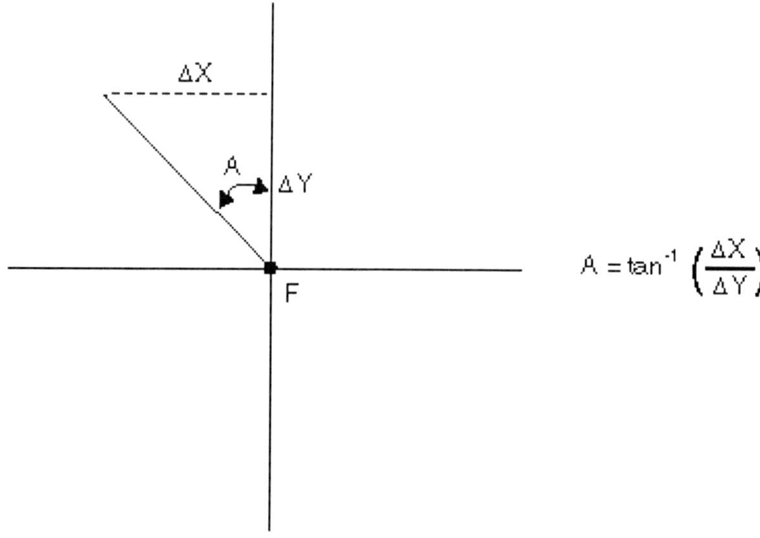

$$A = \tan^{-1}\left(\frac{\Delta X}{\Delta Y}\right)$$

The following calculations were used to determine the rank in Figure A3.4 on page 177:

given

$$R_F = 1E5\left[(Q-1) + \tan\left(\frac{A}{2}\right)\right]$$

since

$$Q = 3$$

then

$$R_F = 1E5\left[2 + \tan\left(\frac{A}{2}\right)\right]$$

Figure A3.4 Calculating Rank in Quadrant 3

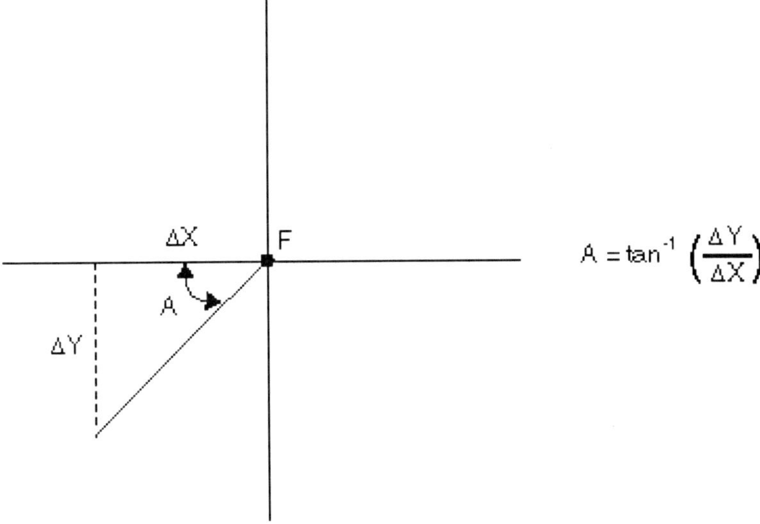

The following calculations were used to determine the rank in Figure A3.5 on page 178:

given

$$R_F = 1E5\left[(Q-1) + \tan\left(\frac{A}{2}\right)\right]$$

since

$$Q = 4$$

then

$$R_F = 1E5\left[3 + \tan\left(\frac{A}{2}\right)\right]$$

Figure A3.5 Calculating Rank in Quadrant 4

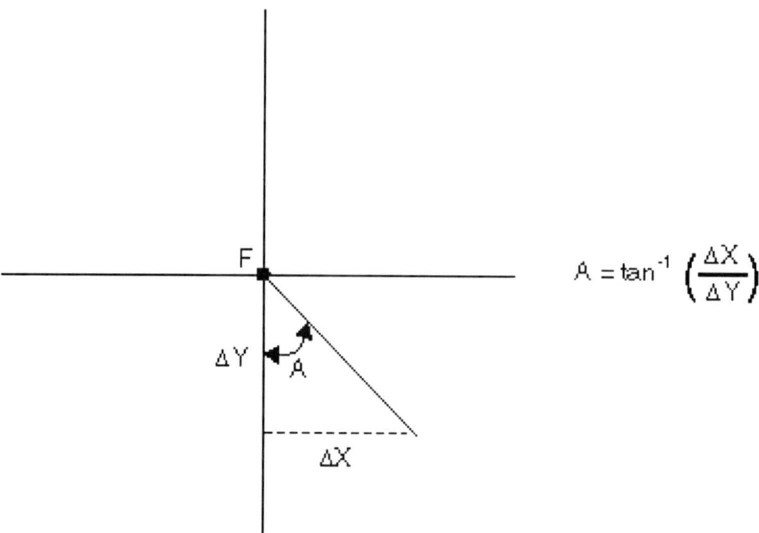

Chain Rank Calculation Examples

The following sections contain information on calculating these values:

- □ chain rank
- □ from-node
- □ to-node

Point Coordinates

The examples in "Example: Calculating From-Node Rank" on page 179 and "Example: Calculating To-Node Rank" on page 180 use the information in the following table.

Table A3.1 Coordinate Values for a Chain with One Detail Point

Point	Description	X	Y
F	from-node of chain	-784533	373266
D	Detail point	-784688	373375
T	To-node of chain	-784559	373498

Variable Definitions

The variables used in the equations have the following definitions:

R_F	is the rank value at the from-node of the chain.
R_T	is the rank value at the to-node of the chain.
A	is the angle from the chain clockwise to the nearest X or Y axis.
ΔX	is the length of a chain segment along the X axis.
ΔY	is the length of a chain segment along the Y axis.

Calculating Chain Rank

The from-node and to-node rank values are expressed as a single number in the form *ffffff.tttttt*. Therefore, the rank for the chain in "Example: Calculating From-Node Rank" on page 179 and "Example: Calculating To-Node Rank" on page 180 is 131641.240034. This is the value of the RANK variable for this chain in the chains data set.

Note: The trigonometric functions for calculating the RANK value in the following sections are in radians. △

Example: Calculating From-Node Rank

The following equations illustrate the steps necessary to calculate the from-node rank: given

$$R_F = 1E5\left\{(Q - 1) + \tan\left[\frac{A}{2}\right]\right\}$$

and

$$A = \tan^{-1}\left(\frac{\Delta Y}{\Delta X}\right)$$

then

$$R_F = 1E5\left\{(Q - 1) + \tan\left[\left(\frac{1}{2}\right)\tan^{-1}\left(\frac{\Delta Y}{\Delta X}\right)\right]\right\}$$

The following example illustrates calculating a from-node rank with the given values: given

$$Q = 2$$

and

$$|X_D - X_F| = |-784688 - (-784533)| = 155$$

and

$$|X_D - X_F| = |-373375 - 373266| = 109$$

then

$$R_F = 1E5\left\{(2 - 1) + \tan\left[\left(\frac{1}{2}\right)\tan^{-1}\left(\frac{109}{155}\right)\right]\right\} = 131,641$$

Figure A3.6 Calculating the From-Node Rank

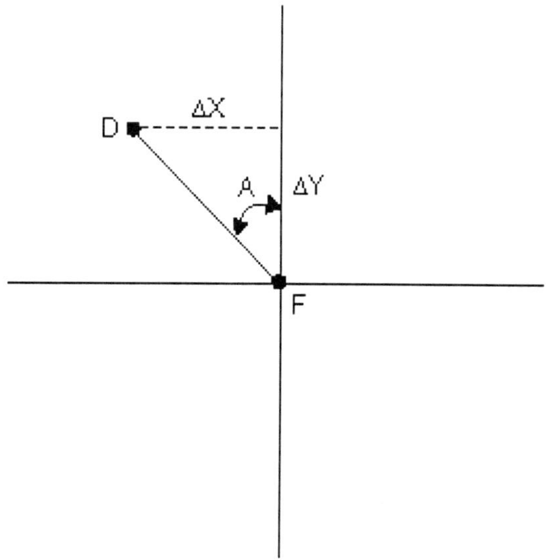

Example: Calculating To-Node Rank

The following equations illustrate the steps necessary to calculate the to-node rank: given

$$R_T = 1E5\left\{(Q - 1) + \tan\left[\frac{A}{2}\right]\right\}$$

and

$$A = \tan^{-1}\left(\frac{\Delta Y}{\Delta X}\right)$$

then

$$R_T = 1E5\left\{(Q - 1) + \tan\left[\left(\frac{1}{2}\right)\tan^{-1}\left(\frac{\Delta Y}{\Delta X}\right)\right]\right\}$$

The following example illustrates calculating a to-node rank with the given values: given

$$Q = 3$$

and

$$|X_D - X_F| = |-784688 - (-784559)| = 129$$

and

$$|X_D - X_F| = |-373375 - 373498| = 123$$

then

$$R_T = 1E5\left\{(3 - 1) + \tan\left[\left(\frac{1}{2}\right)\tan^{-1}\left(\frac{123}{129}\right)\right]\right\} = 240,034$$

Figure A3.7 Calculating the To-Node Rank

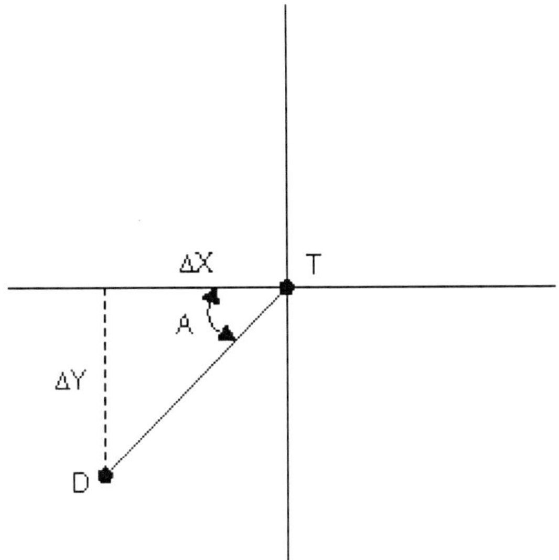

APPENDIX 4

Recommended Reading

Recommended Reading

Here is the recommended reading list for this title:

- □ *SAS/ACCESS for Relational Databases: Reference*
- □ *SAS/FSP Procedures Guide*
- □ *SAS/GRAPH: Reference*
- □ *SAS Language Reference: Concepts*
- □ *SAS Language Reference: Dictionary*
- □ *Base SAS Procedures Guide*

For a complete list of SAS publications, go to **support.sas.com/bookstore**. If you have questions about which titles you need, please contact a SAS Publishing Sales Representative at:

SAS Publishing Sales
SAS Campus Drive
Cary, NC 27513
Telephone: 1-800-727-3228
Fax: 1-919-531-9439
E-mail: **sasbook@sas.com**
Web address: **support.sas.com/bookstore**

Customers outside the United States and Canada, please contact your local SAS office for assistance.

Glossary

ArcInfo
a geographic information system software product that is developed and supported by ESRI.

area
a type of map feature that represents a closed (polygonal) area such as a state or county. Area boundaries are defined by individual line segments. An area can also be the size or extent of the enclosed polygon as calculated by the GIS procedure. See also perimeter.

area layer
a layer that contains the regions, such as census tracts or ZIP code zones, that are part of a map. See also area.

attribute data
values that are associated with features on a map. Attribute data is linked to map features by key variables. Attribute data can include information such as household incomes, population, sales revenue, ages, and so on. See also spatial data and key variable.

chain
a sequence of two or more points in the coordinate space. The end points (that is, the first and last points of the chain) are called nodes. See also detail point and node.

coverage
a subset of the spatial data that is available to a map. For example, a coverage might include the spatial data for a region of a map that is contained in a spatial database.

coverage entry
a SAS catalog entry of type GISCOVER that defines the subset, or coverage, of the spatial data that is available to a map.

data type
an attribute of every column in a table or database. The data type tells the operating system how much physical storage to set aside for the column and specifies what type of data the column will contain. It is similar to the type attribute of SAS variables.

data value
a unit of character or numeric information in a SAS data set. A data value represents one variable in an observation. For example, the variable LASTNAME might contain the data value Smith.

database management system
a software application that enables you to create and manipulate data that is stored in the form of databases. Short form: DBMS.

DBMS
See database management system.

detail point
an intermediate point that delineates the interior segment of a line. Detail points are those points on a line between the from-node and the to-node. See also point, to-node, and from-node.

DLG
Digital Line Graph. A data exchange format for planimetric data. DLG was developed by the United States Geological Survey (USGS).

DXF
a data exchange format that is widely used in computer-aided design (CAD) applications.

Dynamap
spatial (map) data that is developed and supported by Tele Atlas NV.

external file
a file that is created and maintained by a host operating system or by another vendor's software application. SAS can read data from and route output to external files. External files can contain raw data, SAS programming statements, procedure output, or output that was created by the PUT statement. A SAS data set is not an external file. See also fileref.

feature
a man-made or natural physical object such as a river, utility line, school, street, or highway; or an intangible boundary or area such as a sales territory, a census tract, a county boundary, or a state boundary.

feedback area
an area in the GIS Map window that displays information about the map scale, as well as about location coordinates, distance values, and attribute values.

fileref
a name that is temporarily assigned to an external file or to an aggregate storage location such as a directory or a folder. The fileref identifies the file or the storage location to SAS. See also libref.

from-node
the starting coordinates of a line segment on a SAS/GIS map. See also detail point, node, point, and to-node.

generic import file
a file that contains spatial data that you can import by writing a SAS DATA step to convert the data to a SAS/GIS generic form. Once the data is in this generic form, SAS/GIS software can finish the import process.

geocoding
a process for calculating an X and Y coordinate for a street address.

geographic information system
a software application for organizing and analyzing data that can be referenced spatially - that is, data that can be associated with physical locations. Many types of data, such as data from marketing surveys and epidemiological studies, have a spatial aspect. Short form: GIS.

GIS
See geographic information system.

key variable
a variable that is used to associate attribute data with specific features on a map. The key variable has the same data type (character or numeric) and the same data value in the spatial database as in the attribute data set.

label data set
a data set that defines the attributes (location, color, size, and so on) of labels that will be displayed on a map.

layer
a group of features that have the same attribute. For example, all of the lines that are streets, all of the points that are houses, and all of the areas that are census tracts are layers. See also static layer and thematic layer.

layer definition
a WHERE clause that is applied to spatial data in order to specify which features will be displayed in a layer.

layer entry
a SAS catalog entry of type GISLAYER that includes the type of the layer (point, line, or area), as well as a layer definition and information about the graphical characteristics of the layer, such as the line color, the point symbol, or the fill pattern. See also layer definition.

line
in topological terms, a one-dimensional feature that is defined by two zero-dimensional features (points). A line starts at a designated point (the from-node) and ends at a designated point (the to-node), but it can also have intermediate (detail) points. Lines can represent streets, rivers, or boundaries. A line can also be referred to as a chain. See also from-node and to-node.

map
a graphic representation of an area. The area is often a geographic area, but it can also be any other area of any size.

map area
a polygon or group of polygons on a map. For example, states, provinces, and countries are typical map areas. In a map data set, a map area consists of all the observations that have the same values for the identification variable or variables. A map area is sometimes referred to as a unit area.

map data set
a SAS data set that contains information that the GMAP procedure uses to draw a map. Each observation in the data set contains variables whose values are the x, y coordinates of a point on the boundary of a map area. In addition, each observation contains an identification variable whose value identifies the map area that the point belongs to.

map entry
a SAS catalog entry of type GISMAP that contains the layers, links to key variables, the name of the label data set, the name of the coverage entry, legend information, and so on for a map.

Map window
the SAS/GIS window that displays the current map. The Map window enables you to interactively query attribute data and to modify the map.

MapInfo
a GIS software application that is developed and supported by MapInfo Corporation.

node
a point on a map that has connections to one or more chains. See also chain and point.

perimeter
the total length of the sides of a closed polygon. The perimeter value is calculated by the GIS procedure when the AREA option is used. See also area.

point
in topological terms, a zero-dimensional feature that is the base component upon which higher dimensional objects (lines and polygons) are defined. A point can represent a feature such as a house, a store, or a town. See also detail point.

polygon
a closed geometric figure that is bounded by lines or arcs. A polygon can be filled to represent a surface.

SAS catalog
a SAS file that stores many different kinds of information in smaller units called catalog entries. A single SAS catalog can contain several different types of catalog entries.

SAS data set
a file whose contents are in one of the native SAS file formats. There are two types of SAS data sets: SAS data files and SAS data views. SAS data files contain data values in addition to descriptor information that is associated with the data. SAS data views contain only the descriptor information plus other information that is required for retrieving data values from other SAS data sets or from files that are stored in other software vendors' file formats.

SAS library
a collection of one or more files that are recognized by SAS and that are referenced and stored as a unit. Each file is a member of the library.

SAS/GIS software
a SAS software product that provides an interactive windowing environment for analyzing and displaying data in a spatial or geographic context.

SAS/GRAPH software
a SAS software product that analyzes data and that visually represents the relationships between data values as two- and three-dimensional graphs.

shapefile
a format for spatial data files developed by ESRI.

spatial analysis
the process of analyzing data that can be referenced spatially in order to extract or generate new geographical information.

spatial data
coordinates and other information that are used for drawing maps. The maps can include features such as city boundaries, census tract boundaries, streets, schools, and so on. Spatial data is stored in three SAS data sets: the chains, nodes, and details data sets. See also attribute data.

spatial database
a database that contains the following three SAS/GIS data sets: chains, nodes, and details. A spatial database also contains catalog entries that define the information that is needed in order to display a map.

static layer
a layer in which the values of the graphical characteristics (fill color, outline color, line width, and so on) are the same for all features in the layer. See also layer and thematic layer.

thematic layer
a layer in which the graphical characteristics for each feature in the layer are determined by the values of response variables in an associated attribute data set. For example, line widths on a highway layer can represent traffic volumes, and fill colors on an area layer can represent population densities. See also layer and static layer.

TIGER
Topologically Integrated Geographic Encoding and Referencing. A format for map data that was developed by the United States Census Bureau. As of March 2008, the TIGER file format has been superceded by the TIGER shapefile format. See also shapefile.

to-node
the ending coordinates of a line segment on a SAS/GIS map. See also detail point, from-node, node, and point.

tool palette
in the GIS Map window, the collection of icons that represent functions in the interface.

variable
a column in a SAS data set or in a SAS data view. The data values for each variable describe a single characteristic for all observations. Each SAS variable can have the following attributes: name, data type (character or numeric), length, format, informat, and label.

Index

Your Turn

We welcome your feedback.

- If you have comments about this book, please send them to **yourturn@sas.com**. Include the full title and page numbers (if applicable).
- If you have comments about the software, please send them to **suggest@sas.com**.

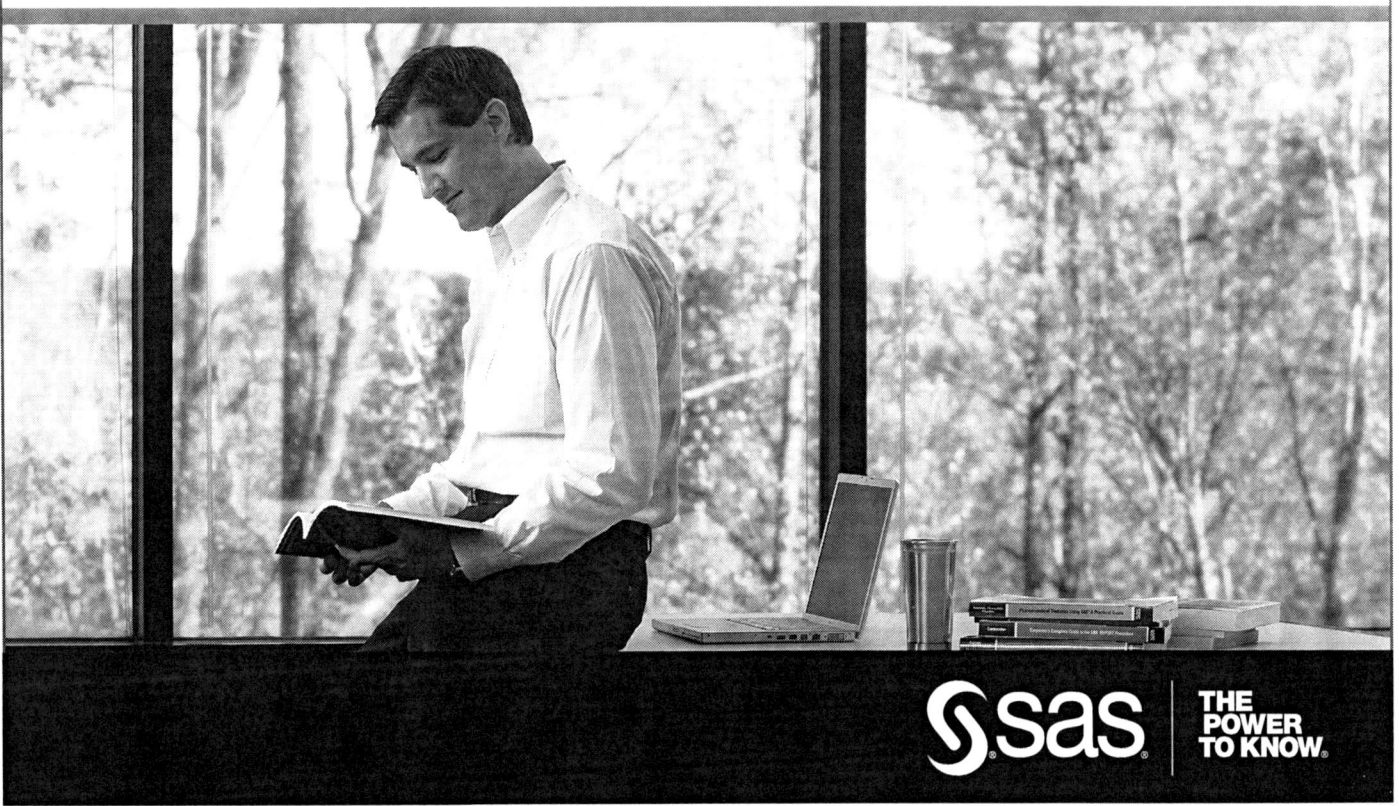

LaVergne, TN USA
27 January 2010
171352LV00001B/77/P

9 781599 948898